Bioremediation of Metals and Inorganic Compounds

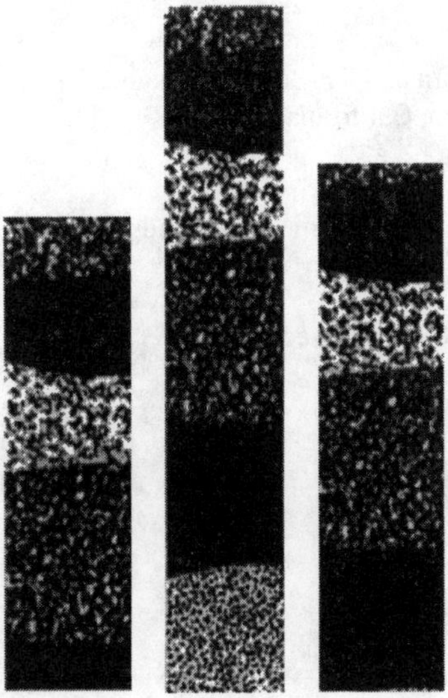

Editors

Andrea Leeson
and Bruce C. Alleman
Battelle

The Fifth International In Situ and
On-Site Bioremediation Symposium

San Diego, California, April 19–22, 1999

BATTELLE PRESS
Columbus • Richland

Library of Congress Cataloging-in-Publication Data

Bioremediation of metals and inorganic compounds / editors, Andrea Leeson and Bruce C. Alleman
p. cm.
Proceedings from the Fifth International In Situ and On-Site Bioremediation Symposium, held April 19–22, 1999, in San Diego, California.
Includes bibliographical references and index.
ISBN 1-57477-077-2 (hardcover : alk. paper)
1. Metal wastes--Biodegradation Congresses. 2. Inorganic compounds--Biodegradation Congresses. 3. Bioremediation Congresses.
I. Leeson, Andrea, 1962– . II. Alleman, Bruce C., 1957– .
III. International Symposium on In Situ and On-Site Bioremediation (5th : 1999 : San Diego, Calif.)
TD196.M4B5723 1999
628.5'2--dc21

99-23329
CIP

Printed in the United States of America

Battelle Press
505 King Avenue
Columbus, Ohio 43201, USA
614-424-6393 or 1-800-451-3543
Fax: 1-614-424-3819
Internet: press@battelle.org
Website: www.battelle.org/bookstore

For information on future environmental conferences, write to:
Battelle
Environmental Restoration Department, Room 10-123B
505 King Avenue
Columbus, Ohio 43201-2693
Phone: 614-424-7604
Fax: 614-424-3667
Website: www.battelle.org/conferences

CONTENTS

Foreword v

Using Artificial Neural Networks to Assess Changes in Microbial Communities. *C.C. Brandt, J.C. Schryver, S.M. Pfiffner, A.V. Palumbo, and S. Macnaughton* 1

Impact of Treatment Additives Used to Reduce Lead Solubility and Microbial Toxicity in Contaminated Soils. *V. Jonioh, J.P. Obbard, and R.R. Stanforth* 7

Chromium Elimination with Microbially Reduced Iron: Redox-Reactive Biobarriers. *R. Gerlach, A. Cunningham, and F. Caccavo* 13

Remediation of a Heavy Metal Plume Using a Reactive Wall. *R. McGregor, D. Blowes, R. Ludwig, E. Pringle, and M. Pomeroy* 19

Assessing Metal Bioavailability in Soils Using Luminescence-Based Microbial Biosensors. *S. Preston, J.M. Hall, E.A.S. Rattray, A.M. Chaudri, S.P. McGrath, K. Killham, and G.I. Paton* 25

Microbiological Treatment of Cyanide, Metals, and Nitrates in Mining Byproducts. *S. Nordwick, M. Canty, R. Hiebert, and L. Thompson* 31

Studies on Biorecovery of Metals from Acid Mine Drainage. *R. Govind, W. Yang, and H.H. Tabak* 37

Use of Vegetable Oil in a Pilot-Scale Denitrifying Biobarrier. *W.J. Hunter* 47

Bioremediation of Nitrate-Contaminated Soils Using Innocuous Oil. *W.J. Hunter* 53

In Situ Denitrification: A Field Demonstration. *H.E. Nuttall, L. Deng, A. Abdelouas, and W. Lutze* 59

Denitrification Kinetics: Model and Comparison to Experimental Data. *Y. Lu, H.E. Nuttall, and W. Lutze* 65

In Situ Autotrophic Denitrification Using Electrodes Under Oligotrophic Conditions. *Y. Sakakibara and J. Kusaka* 73

Combined Microbial-Fe(0) System to Treat Nitrate-Contaminated Groundwater. *T.D. DeJournett and P.J.J. Alvarez* 79

Denitrification of Septic Tank Effluent Using Sulfur/Limestone Processes. *J. Shan and T.C. Zhang* 85

Laboratory and Field Testing of Engineered Denitrification. *S.N. Nelson, K.T. O'Reilly, M.L. Barbonio, and G. Vogt* 91

Removal of Nitrate from Aqueous Streams. *E.A. Stadlbauer, M. Röhricht, H. Happel, B. Weber, J. Eberheim, U. Sauerland, S. Bojanowski, and J. Esch* 97

Bioremediation In Situ of Soils Contaminated with Radioactive Elements. *S.N. Groudev, P.S. Georgiev, I.I. Spasova, and K. Komnitsas* 103

Field-Scale Evaluation of Microbial Mats for Metal Removal in a Funnel-and-Gate Configuration. *P. Phillips, J. Bender, and C.F. Thornton* 109

Long-Term Stability of Uraninite Following In Situ Remediation of U(VI). *A. Abdelouas, W. Lutze, and E.H. Nuttal* 115

Chromate Removal from Contaminated Groundwater Using Indigenous Bacteria. *J.S. McLean, T.J. Beveridge, and D. Phipps* 121

Enhanced Removal of Copper From Contaminated Silt Soil Using Bioelectrokinesis. *G. Maini, A.K. Sharman, S.A. Jackman, C.J. Knowles, and G. Sunderland* 127

Organic Electron Donors for Microbial Cr(VI) Reduction. *J. Hong and G.W. Sewell* 135

In Situ Microbial Volatilization of Selenium in Soils: A Case History. *P.T. Zawislanski, S.M. Benson, G.R. Jayaweera, L. Wu, and W.T. Frankenberger* 141

Biological Treatment of Perchlorate-Contaminated Waters. *B.E. Logan, K. Kim, J. Miller, P. Mulvaney, and R. Unz* 147

Author Index 153

Keyword Index 181

FOREWORD

The Fifth International In Situ and On-Site Bioremediation Symposium was held in San Diego, California, April 19–22, 1999. The program included approximately 600 platform and poster presentations, encompassing laboratory, bench-scale, and full-scale field studies being conducted worldwide on a variety of bioremediation and supporting technologies used for a wide range of contaminants.

The author of each presentation accepted for the program was invited to prepare a six-page paper, formatted according to specifications provided by the Symposium Organizing Committee. Approximately 400 such technical notes were received. The editors conducted a review of all papers. Ultimately, 389 papers were accepted for publication and assembled into the following eight volumes:

Natural Attenuation of Chlorinated Solvents, Petroleum Hydrocarbons, and Other Organic Compounds – Volume 5(1)
Engineered Approaches for In Situ Bioremediation of Chlorinated Solvent Contamination – Volume 5(2)
In Situ Bioremediation of Petroleum Hydrocarbon and Other Organic Compounds – Volume 5(3)
Bioremediation of Metals and Inorganic Compounds – Volume 5(4)
Bioreactor and Ex Situ Biological Treatment Technologies – Volume 5(5)
Phytoremediation and Innovative Strategies for Specialized Remedial Applications – Volume 5(6)
Bioremediation of Nitroaromatic and Haloaromatic Compounds – Volume 5(7)
Bioremediation Technologies for Polycyclic Aromatic Hydrocarbon Compounds – Volume 5(8)

Each volume contains comprehensive keyword and author indices to the entire set.

This volume deals with bioremediation as an approach to remediating metals and inorganic compound contamination. Bioremediation has been a problematic alternative for inorganics. Unlike organic contaminants, which often can be broken down by biological processes into relatively harmless metabolites and byproducts, metals are elemental. Further, metals and their salts often have an inhibitory effect on biological activity. However, despite these potential pitfalls, there has been progress recently in applying bioremediation technologies to metals and inorganics. This volume encompasses topics such as lead solubility reduction, chromium reduction, denitrification, volatilization of selenium in soils, metals recovery from acid mine drainage, and even the possibility of applying artificial neural network technology to aid in bioremediation.

We would like to thank the Battelle staff who assembled the eight volumes and prepared them for printing. Carol Young, Lori Helsel, Loretta Bahn, Gina Melaragno, Timothy Lundgren, Tom Wilk, and Lynn Copley-Graves spent many hours on production tasks—developing the detailed format specifications sent to each author; examining each technical note to ensure that it met basic page layout requirements and making adjustments when necessary; assembling the volumes; applying headers and page numbers; compiling the tables of contents and author and keyword indices, and performing a final check of the pages before submitting them to the publisher. Joseph Sheldrick, manager of Battelle Press, provided valuable production-planning advice and coordinated with the printer; he and Gar Dingess designed the covers.

The Bioremediation Symposium is sponsored and organized by Battelle Memorial Institute, with the assistance of a number of environmental remediation organizations. In 1999, the following co-sponsors made financial contributions toward the Symposium:

- Celtic Technologies
- Gas Research Institute (GRI)
- IT Group, Inc.
- Parsons Engineering Science, Inc.
- U.S. Microbics, Inc.
- U.S. Naval Facilities Engineering Command
- Waste Management, Inc.

Additional participating organizations assisted with distribution of information about the Symposium:

- Ajou University, College of Engineering
- American Petroleum Institute
- Asian Institute of Technology
- Conor Pacific Environmental Technologies, Inc.
- Mitsubishi Corporation
- National Center for Integrated Bioremediation Research & Development (University of Michigan)
- U.S. Air Force Center for Environmental Excellence
- U.S. Air Force Research Laboratory Air Base and Environmental Technology Division
- U.S. Environmental Protection Agency
- Western Region Hazardous Substance Research Center (Stanford University and Oregon State University)

The materials in these volumes represent the authors' results and interpretations. The support of the Symposium provided by Battelle, the co-sponsors, and the participating organizations should not be construed as their endorsement of the content of these volumes.

Andrea Leeson and Bruce Alleman, Battelle
1999 Bioremediation Symposium Co-Chairs

USING ARTIFICIAL NEURAL NETWORKS TO ASSESS CHANGES IN MICROBIAL COMMUNITIES

Craig C. Brandt, (Oak Ridge National Laboratory, Oak Ridge, Tennessee)
Jack C. Schryver (Oak Ridge National Laboratory, Oak Ridge, Tennessee)
Susan M. Pfiffner, (University of Tennessee, Knoxville, Tennessee)
Anthony V. Palumbo (Oak Ridge National Laboratory, Oak Ridge, Tennessee)
Sarah Macnaughton, (University of Tennessee, Knoxville, Tennessee)

ABSTRACT: We evaluated artificial neural networks (ANNs) as a technique for assessing changes in soil microbial communities following exposure to metals. We analyzed signature lipid biomarker (SLB) data collected from two soil microcosm experiments using traditional statistical techniques and ANN. Two phases of data analysis were done; pattern recognition and prediction. In general, the ANNs were better able to detect patterns and relationships in the SLB data than were the traditional statistical techniques.

INTRODUCTION

A major challenge in the implementation of in situ bioremediation technologies is understanding the indigenous microbial community structure and how this structure is affected by environmental conditions. Understanding the microbial community structure and the environmental factors that control it would assist in optimizing those factors that offer the best opportunity for control of bioremediation. For example, assessing the changes in the community structure following the addition of nutrient amendments is critical in monitoring the effectiveness of bioremediation.

Microbial communities in soils have often been characterized by the analysis of signature lipid biomarkers (SLBs) (e.g., Tunlid and White, 1992). SLBs are known to respond to changes in environment conditions such as temperature and nutrients. However, these changes are often complex, nonlinear, and not readily amenable to traditional statistical analyses.

In this study we used artificial neural networks (ANNs) to investigate the complex relationships among SLBs. ANNs are nonlinear, nonparametric analysis methods that can learn from experience to improve their performance. These methods are tolerant of missing or noisy data and can be used for rapid analysis of new data sets.

An ANN consists of many simple processing elements called nodes. Nodes are interconnected by means of directed communication links, each with an associated weight. Each node has an internal state, called its activation level, which is a function of the inputs it receives. A node sends its activation as a signal to the other nodes with which it is connected (Fausett, 1994). Figure 1 illustrates the architecture of a simple ANN consisting of five input parameters, eight nodes, and five output parameters.

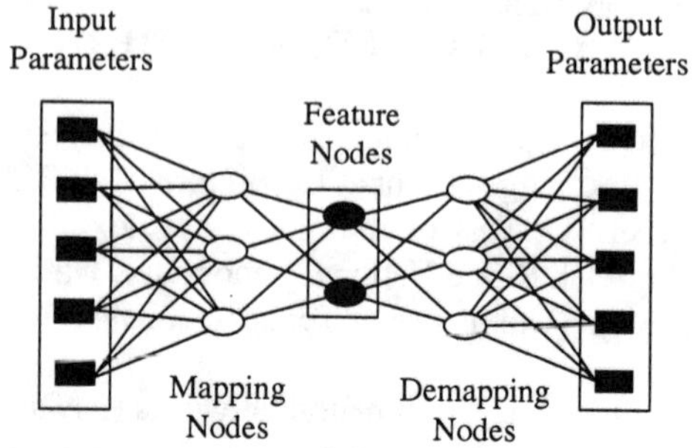

FIGURE 1. Example of a simple artificial neural network consisting of five input parameters, three mapping nodes, two feature nodes, three demapping nodes, and five output parameters. Solid lines represent directed communication links between components of the ANN.

The weights associated with the communication links in an ANN represent information used by the net to solve a problem. An ANN is trained by presenting it with a paired set of input and output patterns. The ANN adjusts its weights in order to minimize the difference between the output parameters calculated by the net and the actual output pattern.

MATERIAL AND METHODS

For this study, we analyzed SLB data obtained from two soil microcosm experiments that were designed to determine the effects of metal concentration of the microbial community structure. Two phases of data analysis were conducted; pattern recognition and prediction. In the pattern recognition phase, the SLB data were analyzed without regard to the applied experimental treatments (e.g., metal concentration). In the prediction phase, we used the SLB data to predict the experimental treatment associated with each sample. In both phases, the training data were analyzed first, and the resulting models were then applied to the validation data to test the robustness of the analysis.

Microcosms. Agricultural topsoil collected from the University of Tennessee agricultural experiment station was used in both microcosm experiments. In the first experiment, the microcosms were exposed to 0 and 500 ppm of a metal mixture that contained cadmium, strontium, cobalt, and cesium (Flemming et al., 1998). The second experiment used 0, 100, and 250 ppm concentrations of the same metal mixture (Macnaughton et al., 1998). Triplicate microcosms from each metal treatment were sacrificed at 0, 7, 14, 28, and 56 days.

The sacrificed microcosms were analyzed for SLBs using the procedure described in White and Ringelberg (1998). Total lipids were extracted from the microcosm samples using a modified chloroform and methanol solution. Silicic acid column chromatography was used to separate the total lipid extract into general lipid classes. Methyl esters of the SLBs were prepared using alkaline methanolysis, and these esters were identified and quantified using gas chromatography/mass spectrometry.

Those SLBs that were at or below the detection limit in all samples were dropped from the analysis. The resulting profiles contained 43 SLBs. The data from the two experiments were combined into a single data set, transformed to a 0-1 scale, and divided into three subsets prior to analysis. The subsets were recombined to create a training data set consisting of two-thirds of the data and a validation data set consisting of the remaining one-third of data.

Pattern Recognition. In the pattern recognition phase, we used SAS to conduct a two factor principal components analysis (PCA) on the training data (SAS, 1989). The factor coordinates for each sample were input into a hierarchical cluster analysis procedure to identify groupings of samples (SAS, 1989). The coordinates and associated cluster assignments were plotted.

We used the SLBs to train an autoassociative ANN using the Stuttgart Neural Network Simulator (Zell et al., 1994). The ANN architecture consisted of 43 input SLBs, six mapping nodes, two feature nodes, seven demapping nodes, and 43 SLBs. The coordinates of the ANN's feature nodes were analyzed by hierarchical cluster analysis, and the coordinates and cluster assignments were plotted (SAS, 1989).

Prediction. For the prediction phase, we used SAS to conduct a multivariate linear regression (MLR) analysis of the training data (SAS, 1989). The SLBs were the independent or predictor variables and the treatments (experiment, metal concentration, and time) were the dependent or endpoint variables. The coefficients of the MLR model developed from the training data were used to analyze the validation data.

We used the Stuttgart Neural Network Simulator (Zell et al., 1994) to create a feed-forward ANN to predict experiment, metal concentration, and time from the SLBs. The ANN architecture consisted of 43 SLBs, seven feature nodes, and three endpoint nodes corresponding to experiment, metal concentration, and time. The ANN model developed from the training data was used to analyze the validation data.

RESULTS

Pattern Recognition. The two-factor PCA explained 45.2 percent of the variance in the training data set (Figure 2). The enclosed samples represent groups identified by the hierarchical cluster analysis. In general, the PCA separated the results of the first (groups 5 - 8) and second (groups 1 - 4) experiments. Within the first experiment, the high metal samples (group 5) are separate from the no metal samples after 7 days (group 6). A similar separation of the metal treated samples (group 1) from the untreated samples (groups 3 and 4) is also seen in second experiment.

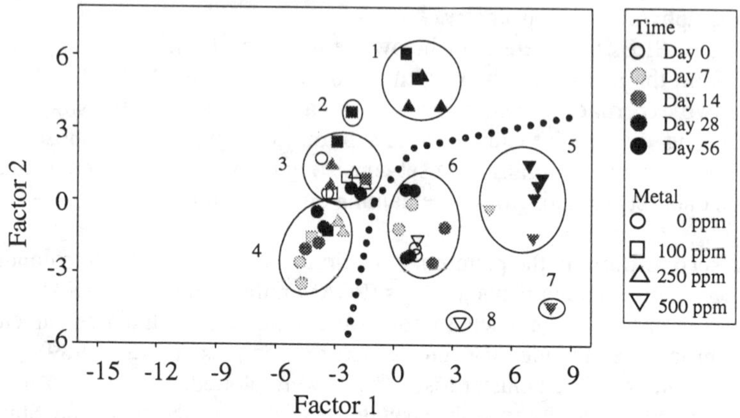

FIGURE 2. Plot of factor coordinates from the PCA analysis of the microcosm training data. Sample groupings were determined by a hierarchical cluster analysis of the node coordinates.

The two-node autoassociative ANN explained 77.6 percent of the variance in the training data set (Figure 3). The enclosed samples represent groups identified by the hierarchical cluster analysis. Like the PCA, the ANN separated the results of the first (groups 1 - 6) and second (groups 7 - 9) experiments. For the first experiment, the untreated samples (groups 1, 2, 4, and 6) are distinct from treated samples (groups 3 and 4). Within the second experiment, the treated samples (group 9) are distinct from the untreated samples (groups 7 and 8).

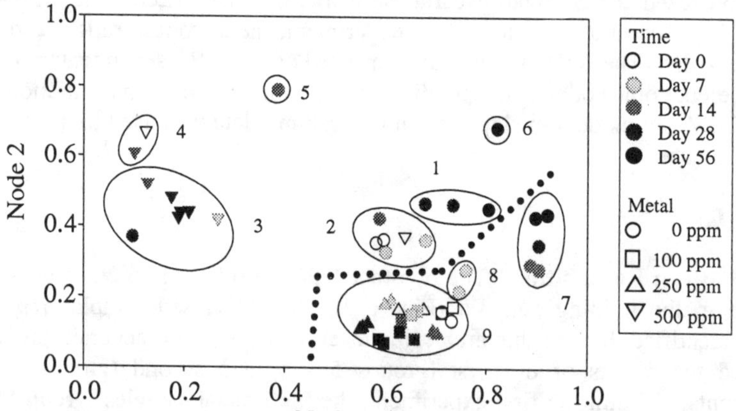

FIGURE 3. Plot of feature node coordinates from the autoassociative ANN analysis of the microcosm training data. Sample groupings were determined by a hierarchical cluster analysis of the node coordinates.

Prediction. Overall, the MLR analysis explained 93.0 percent of the variance in the training data compared to 97.8 percent for the feed-forward ANN (Table 1). The ANN outperformed the MLR in predicting experiment and metal concentration, but the MLR did better at predicting the time endpoint.

The MLR model developed for the training data explained 82.6 percent of the variance when applied to the validation data (Table 1). For these data, the MLR explained the most variance in experiment, followed by metal and time. The feed-forward ANN developed for the training data explained 95.0 percent of the variance in the validation data. The ANN explained the most variance in experiment, followed by time and metal.

TABLE 1. Percent of variance explained by multivariate linear regression (MLR) and artificial neural network (ANN) analyses of the training and validation microcosm data.

	Training Data		Validation Data	
Endpoint	**MLR**	**ANN**	**MLR**	**ANN**
Experiment	83.7	99.9	94.2	99.7
Metal	94.4	95.4	77.3	88.2
Time	97.7	96.4	67.0	93.7
OVERALL	93.0	97.8	82.6	95.0

In general, the ANNs explained more of the variance in the training data than did the traditional statistical techniques. While the overall results for the autoassociative ANN and PCA were similar, the ANN identified more distinct clusters in the feature coordinates than did the PCA. The feed-forward ANN also generalized to the validation data better than did the MLR technique. Both data analysis phases uncovered experimental differences in the SLB data that we believe are caused by changes in the microbial communities due to storage of the soil between experiments (Findlay et al., 1985).

Based on the results of this study, we believe that ANNs offer a more robust approach for analyzing and interpreting complex data sets, and generalizing to new data sets. ANNs can be used as a tool for assessing microbial community structure changes as a result of pollution. We hope these tools will be of particular relevance in the assessment of natural and accelerated bioremediation in soil.

ACKNOWLEDGMENTS

This research is supported by the U.S. Department of Energy Natural and Accelerated Bioremediation Research (NABIR) program. We would like to thank NABIR Science Team Leader Dr. D.C. White and personnel at the Center for Environmental Biotechnology at the University of Tennessee who provided the SLB data used in this research. Oak Ridge National Laboratory is managed by Lockheed Martin Energy Research, for the U.S. Department of Energy under contract DE-AC05-96OR22464.

REFERENCES

Fausett, L. V. 1994. *Fundamentals of Neural Networks.* Prentice Hall, NJ.

Findlay, R. V., P. C. Pollard, D. J. W. Moriarty, and D. C. White. 1985. Quantitative determination of microbial activity in community nutritional status in estuarine sediments: Evidence for a disturbance artifact." *Cand. J. Microbiol.* 31: 493-498.

Flemming, C. A., K. T. Leung, A. Peacock, S. Macnaughton, and D. C. White. 1999. "Changes in microbial community lipid, DNA and RNA biomarkers in soil contaminated with cadmium, cobalt, cesium, and strontium." (in preparation).

Macnaughton, S., J. R. Stephen, Y. J. Chang, A. Peacock, C. A. Flemming, K. T. Leung, and D. C. White. 1998. "Characterization of metal resistant soil eubacteria by PCR DGGE with isolation of resistant strains." *Cand. J. Microbiol.* (submitted).

SAS Institute, Inc. 1989. *SAS/STAT User's Guide, Version 6, Fourth Edition, Vols. 1 and 2.* SAS Institute, Inc., Cary, NC.

Tunlid, A. and D. C. White. 1992. "Biochemical analysis of biomass, community structure, nutritional status, and metabolic activity of microbial communities in soils." In G. Stotzky and J. M. Bollag (Eds.), *Soil Biochemistry*, pp. 229-262. Marcel Dekker, Inc., New York, NY.

White, D. C. and D. B. Ringelberg. 1998. "Signature lipid biomarker analysis." In R. S. Burlage, R. Atlas, D. Stahl, G. Geesey, and G. Sayler (Eds.), *Techniques in Microbial Ecology*, pp. 255-272. Oxford University Press, Inc., New York, NY.

Zell, A., N. Mache, R. Hubner, G. Mamier, M. Vogt, M. Schmalzl, and K. U. Herrmann. 1994. "SNNS (Stuttgart Neural Network Simulator)." In J. Skrzypek (Ed.), *Neural Network Simulation Environments*, pp. 165-186. Kluwer Academic Publishers, Assinippi Park, MA.

IMPACT OF TREATMENT ADDITIVES USED TO REDUCE LEAD SOLUBILITY AND MICROBIAL TOXICITY IN CONTAMINATED SOILS

Viola Jonioh, Jeffrey P. Obbard, and Robert R. Stanforth
Department of Chemical and Environmental Engineering,
National University of Singapore, Singapore

ABSTRACT: A range of chemical soil additives was tested to determine their effectiveness in reducing lead leachability and microbial toxicity using the Microtox® bioassay. Calcium carbonate was found to be the best additive for the treatment of a military rifle range soil contaminated with lead and other heavy metals. Microbial toxicity and lead leachability were inversely related to the dosage of calcium carbonate added to the soil, thereby indicating the potential of this additive to minimize adverse effects on the soil microbial ecosystem when treating contaminated sites.

INTRODUCTION

A laboratory study was undertaken to evaluate the potential for treating subsurface soil contamination from a firing range that contains heavy metals, in particular lead, using the soil-treatment additives commonly deployed in the USA. Examples of additives used to immobilize lead and reduce the leachable lead concentration below the United States EPA criterion value for lead of 5 mg/L (as based on the Toxicity Characteristic Leaching Procedure (TCLP) test), are shown in Table 1 (Stanforth et al., 1990).

TABLE 1. Available Additives for Treatment of Heavy Metal Contaminated Soils

Additive	
Metallic iron	Phosphate (triple super phosphate)
Calcium hydroxide	Calcium dihydrogen phosphate
Magnesium hydroxide	Calcium carbonate
Dithiocarbomate	Cement

Although the treatment of metal-contaminated soil using treatment additives is well established (Ruby et al., 1994; Peters and Shem, 1992; Stanforth et al., 1990; Ma et al., 1993), reports describing the impact of treatment additives (and dosages) on indigenous soil microorganisms are not known to the authors. The soil microbial ecosystem is functionally complex and contains key groups of microorganisms that play an integral role in maintaining soil fertility in relation to plant nutrition. Key microbial processes include decomposition of organic substrates into mineral constituents available for plant uptake (e.g. N, P, K), and symbiotic relationships (e.g. nitrogen fixation) between microbial symbionts and the host-plant (Alexander, 1977; Domsch, 1984). Potential ecotoxicological effects caused by additives to treat contaminated soil are therefore of concern.

Objectives. The first objective of this investigation was to assess the potential toxicity of several treatment additives (and selected dosages) on soil microorganisms using the Microtox® technique as a microbial bioassay in a preliminary screening experiment using the firing range soil. Microtox® uses a photometric technique that measures the light emission by a bioluminescent bacterium (*Vibrio fischeri*) in response to a chemical exposure to a toxicant. The technique has been widely used on a range of environmental media (e.g. soils, sediments, water) to assess the potential ecotoxicological effects of both organic and inorganic contaminants (Lee et al., 1995; Rönpagel et al., 1995; Cook and Wells, 1996). The second objective was to examine the effect of altering the dosage of the most favorable additive on soil lead leachability and Microtox® response. The overall aim of the investigation was to identify a favorable treatment additive that resulted in both the immobilization of lead and had least potential impact on soil microorganisms at the firing range site.

Site Description. The firing range is approximately 10 hectares in area and has been in service for over fifty years. Soils comprise decomposed granites with iron and aluminum oxides on the surface, with elevated levels of heavy metals, particularly lead, as a result of the discharge and disposal of spent ammunition and casing materials. Weathered soil from a contaminated earth-mound 'backstop', is currently being washed and deposited by rainfall run-off to an adjacent delta area which contains vegetation showing signs of phytotoxic stress (i.e. stunted growth, leaf chlorosis). Hence, the site would benefit from a soil remediation program to reduce the bioavailability of toxic heavy metals and to alleviate impacts on the vegetation.

MATERIALS AND METHODS

Field Sampling. Soil samples 0-0.5 ft (0-15 cm) were collected from three separate locations within the delta area. A further sample was collected (0-15 cm) from the backstop. These samples were used for the preliminary screening and dosage experiments respectively (see below).

Soil Physicochemical Analysis. Soil physicochemical properties were analyzed in accordance with the ASTM and BS methods. Heavy metal concentrations were determined using inductively coupled plasma analysis. All analysis was completed in triplicate.

Preliminary Screening Experiments. The soil samples from the delta area were combined to form a composite sample for this experiment. This sample was treated with additives at the dosages listed in Table 2. The purpose of the experiment was to determine the most favorable additive with respect to lead immobilization and potential impact upon soil microorganisms, as based on the toxicological response of Microtox® (see below).

TABLE 2. Additives and selected dosages – preliminary screening experiment

Additive	Dosage (% Weight Basis)
1. Triple Super Phosphate (TSP)	4
2. TSP/Magnesium oxide	4
3. Calcium carbonate	15
4. Calcium hydroxide	10
5. Cement	15
6. Dithiocarbamate	7.5

Triplicate sub-samples were subjected to leaching tests to evaluate the leaching potential of lead under neutral pH conditions. The USEPA's TCLP leaching procedure (SW846 Method 1311) was used, as modified by using distilled water instead of acetic acid solution (to mimic rainfall leaching at the site), a lesser amount of soil and solution and eliminating the leachate digestion step. Testing consisted of vigorous mixing of 100g of the soil sample with additives at the different dosages (% on wet weight basis) in 500 mL plastic Nalgene bottles. This was immediately followed by water leach tests on each sample, where a 1:20 ratio of sample (g) to deionized water (mL) were shaken in acid-washed HDPE bottles with an end-over-end tumbler for 18 hours. The leachates were then filtered through a 0.45μ PTFE membrane filter, prior to toxicity and heavy metal analysis.

Dosage Experiment: As a result of the preliminary screening, the most favorable additive, which had the least impact on the test bacterium, was selected for further study to determine the effect of dosage on toxicity and lead immobilization. The backstop soil was used for this experiment. A range of dosages of the additive (i.e. 0, 1, 2.5, 5 and 10% on a soil wet weight basis) were mixed with the soil sample and treated and analyzed as specified above.

Microtox® Basic Test: Rehydrated bacteria were incubated in the liquid sample at 15°C and four half-serial dilutions of the sample for 10 minutes. The toxicant concentration (% of sample) at which a 50% light loss occurs over 10 minutes exposure time of the bacterium is automatically calculated and reported as the EC50 (effective concentration for 50% light loss). A low EC50 result is indicative of a high toxicological response.

RESULTS AND DISCUSSION

Soil Physicochemical Analysis. Mean lead concentration in the composite delta soils was 2,403 mg/kg, and in the backstop soil 29,783 mg/kg. Soil pH was in the range of pH 4.5-6.1 for delta soils and pH 5.3 for the backstop soil. The soil composition in the delta soils comprised mostly sand (66-77%), with lesser amounts of silt (18-27%) and clay (2-7%). The backstop soil was predominantly sand (90.7%), with lesser amounts of silt (7.3%) and clay (2%).

Preliminary Screening Experiment. Table 3 gives the EC50 values and the concentration of lead in water leachates for the composite delta soil sample, and the additives and dosages tested. The results indicate that all additives used, with the exception of calcium hydroxide, cement and dithiocarbamate, are capable of reducing the lead leaching potential in the soil. However, calcium carbonate was found to result in the least impact on the EC50 (i.e. 78.86%), as measured by the Microtox® bioassay and resulted in a 91.50% mean reduction in the concentration of lead in the leachate samples relative to the untreated control, at a dosage rate of 10%. Overall, the results show that although several of the chemical additives tested are effective in reducing the lead leaching potential in soil, some may have a potentially negative effect on the soil microorganisms including triple super phosphate (TSP), TSP/magnesium oxide, calcium hydroxide, cement and dithiocarbamate. Calcium carbonate therefore appears to be the most suitable additive for the firing range soils due to its effectiveness in reducing lead leaching potential as well as having least potential impact on the soil microorganisms.

TABLE 3. Toxicity and Metal Leaching of Soil Treated with Various Chemical Additives - Preliminary Screening Results (Delta Composite Soil)

Additive	Dosage (%Weight Basis)	EC50 (%)	Pb leached (mg/L)
1. Untreated soil	-	65.85	0.035
2. Triple super phosphate	4	18.87	0.002
(TSP)	4	12.84	0.001
3. TSP/magnesium oxide	10	78.86	0.003
4. Calcium carbonate	10	0.54	30.07
5. Calcium hydroxide	15	0.45	3.79
6. Cement	7.5	0.35	0.16
7. Dithiocarbamate			

Dosage Experiment. The results of the treatment of backstop soil with calcium carbonate at the dosages of 0, 1, 2.5, 5 and 10% (based on wet weight basis), are given in Table 4. The dose-response relationship indicates an increase in both the lead leachate concentration and sample toxicity as the calcium carbonate dosage is decreased from 10 to 1%. Other metals analyzed show a similar relationship (Table 4). A decrease in the pH of leachates correlated with a decrease in calcium carbonate dosages indicates that the concentration of lead solubilized by the water extract was pH dependent.

It is not possible to determine from these results whether the observed reduction in toxicity was a response to the increased additive concentration (and effect on pH) or due to the reduction in lead concentration on leachate samples.

TABLE 4. Toxicity and Metal Leaching Potential of Soil Treated with Different Dosages of Calcium Carbonate (Backstop Soil)

Dosage %	pH	EC50 (%)	Pb (mg/L)	Cu (mg/L)	Zn (mg/L)	Ni (mg/L)	Cd mg/L)
0	6.35	14.17	1.05	0.12	2.10	0.04	0.33
1	8.35	46.17	0.32	0.09	0.12	0.01	0.15
2.5	8.40	41.19	0.22	0.08	0.16	0.01	0.11
5	8.48	55.78	0.17	0.06	0.07	-	0.07
10	8.59	74.87	0.11	0.05	0.03	-	0.06

CONCLUSION

Based on the analytical results, calcium carbonate appears to be the most favorable additive for the firing range soils with respect to its ability to reduce lead leachability and impact on the test bacterium. Of all the additives tested, calcium carbonate is most likely to achieve effective remediation in terms of heavy metal leachability and the protection of the soil microbial ecosystem.

A more detailed study involving the treatment of backstop soil with different dosages of calcium carbonate revealed a general increase in both the lead leachability and toxicity with decreasing dosages from 10% to 1%. Field trials are being conducted to determine the optimal dosage of calcium carbonate with respect to TCLP lead leachability, toxicological response and long term weathering effects on the soil additive. This will form the basis of a site remediation strategy for the site.

REFERENCES

Alexander, M. 1977. *Introduction to Soil Microbiology.* John Wiley & Sons, New York, NY.

Cook, N.H., and Wells, P.G. 1996. "Toxicity of Halifax Harbour Sediments: An Evaluation of the Microtox® Solid-Phase Test." *Water Quality Research Journal of Canada.* 31(4): 673-708.

Domsch, K.H. 1984. "Effects of Pesticides and Heavy Metals on Biological Processes in Soil." *Plant and Soil.* 76:367-378.

Lee, K., Siron, R., Tremblay, G.H., and Lavoie, J. 1995. "Application of the Microtox Solid-Phase Test to Monitor the Effectiveness of Bioremediation Strategies." *Canadian Technical Report of Fisheries & Aquatic Sciences.* 0(2050): 152-156.

Ma, Q.Y., Traina, S.J., and Logan, T.J. 1993. "In Situ Lead Immobilization by Apatite." *Environmental Science Technology.* 27(9): 1803-1810.

Peters, R.W., and Shem, L. 1992. "Use of Chelating Agents for Remediation of Heavy Metal Contaminated Soil." *Environmental Remediation: Removing Organic and Metal Ion Pollutants.* Washington, DC.

Rönpagel, K., Lib, W., and Ahlf, W. 1995. "Microbial Bioassays to Assess the Toxicity of Solid-Associated Contaminants." *Ecotoxicology and Environmental Safety.* 3: 99-103.

Ruby, M.V., Davis, A., and Nicholson, A. 1994. "In Situ Formation of Lead Phosphates in Soils as a Method to Immobilize Lead." *Environmental Science Technology.* 28(4): 646-654.

Stanforth, R.R., Chowdhury, A.K., and Stolzenberg, T.R. 1990. "Evaluation of Treatment Alternatives for an EP-Toxic Grey Iron Foundry Emission Control Waste." *ACS Symposium on Emerging Technologies for Hazardous Waste Treatment.* Atlantic City, NJ.

USEPA 1986 (and updates). Test Methods for Evaluating Solid Wastes. SW-846 Office of Solid Waste and Emergency Response, USEPA Washington, DC.

CHROMIUM ELIMINATION WITH MICROBIALLY REDUCED IRON – REDOX-REACTIVE BIOBARRIERS

Robin Gerlach and Al Cunningham (Center for Biofilm Engineering, Bozeman, Montana)
Frank Caccavo Jr. (University of New Hampshire, Durham, New Hampshire)

ABSTRACT: The research presented herein addresses the need to explore strategies for the cost-effective development of permeable reactive barriers in the deep subsurface. The metabolic capabilities of Dissimilatory Metal-Reducing Bacteria (DMRB) appear to have the potential to create zones of reduced indigenous metals (e.g. ferrous iron) in the path of a groundwater contaminant plume forming redox-reactive barriers. Microbially reduced iron can reductively dechlorinate organic pollutants, reduce nitroaromatics, and/or precipitate heavy metals and radionuclides from contaminated waters. Starvation of bacteria has shown to enhance bacterial transport and distribution through porous media, thus providing a possible means for the delivery of bacteria into the subsurface. To demonstrate the potential of starved bacteria for *in situ* metal-reduction, starved bacteria were resuscitated with surface associated iron minerals in batch and column experiments. In both batch and column studies, the microbially produced surface-associated Fe(II) reduced Cr(VI), a common groundwater contaminant, to Cr(III). Cr(III) precipitated on the existing surfaces, forming stable end products and thereby eliminating the chromium compounds from the water phase.

INTRODUCTION

The heavy metal contamination of groundwater can occur together with chlorinated solvents since metal plating facilities often use chlorinated solvents as degreasing agents before the plating process. Also, some U.S. Department of Energy sites are contaminated with complex mixtures of heavy metals, radionuclides, and chlorinated solvents. The remediation of these sites presents a challenge due to the mix of contaminants.

Zero-valent iron based redox-reactive subsurface barriers have been successfully employed for the elimination of heavy metals and chlorinated solvents from groundwaters (Cantrell et al., 1995; Fortner, 1995). However, the costs of establishing such a zero-valent iron based redox-reactive subsurface barrier by excavation and placement of zero-valent iron filings into the aquifer can be prohibitively high depending on the necessary depth of the barrier (Scott et al., 1998; Fortner, 1995).

Ferrous iron (Fe(II)), produced from indigenous ferric iron (Fe(III)), has shown to also mediate the reactions described above leading to the dechlorination of chlorinated solvents, and the precipitation of heavy metals and radionuclides from contaminated groundwater (Scott et al., 1998; Buerge and Hug, 1997; Gorby et al., 1994).

Recently, a technology involving the injection of a chemical reductant (dithionite) to reduce indigenous ferric iron producing redox-reactive ferrous iron was successfully demonstrated (Scott et al., 1998). The use of chemical reductants however, could become very expensive depending on the subsurface conditions. If other reducible compounds are present besides ferric iron (e.g. humics), the amount of chemical required to reduce the indigenous ferric iron could increase significantly without contributing to the desired contaminant reduction.

Therefore, we are currently evaluating the feasibility of using dissimilatory metal-reducing bacteria (DMRB) to selectively reduce indigenous metals, such as ferric iron. DMRB have been shown to facilitate degradation of chlorinated organics and the precipitation of heavy metals (Caccavo et al., 1996; Caccavo and Olson, 1995; Gorby et al., 1994). DMRB can directly reduce these contaminants enzymatically or alternatively DMRB can enzymatically generate ferrous iron from naturally occurring ferric iron minerals. The ferrous iron can in turn chemically dechlorinate chlorinated organics, and reductively immobilize metals and radionuclides (e.g. Cr^{6+}, U^{6+}).

The DMRB metabolism could be used to generate a redox-reactive biobarrier *in situ*. Such a barrier could be generated by stimulating indigenous DMRB populations or inoculating contaminated environments with pre-adapted or genetically engineered DMRB. The economic feasibility of bioaugmenting large volumes of soil *in situ* depends on the ability to transport DMRB over large distances in aquifers. Starvation has been shown to significantly enhance bacterial transport and distribution of bacteria through porous media (Sharp et al. 1999) and holds promise for a field scale application.

The objective of this work was to demonstrate that starved DMRB can be resuscitated and grown on surface associated iron and that the microbially produced Fe(II) provides sufficient reducing power for the reduction of Cr(VI) to Cr(III). The distribution of Cr-species within the experimental systems was also determined.

MATERIAL AND METHODS

Preparation of Starved Cells. Cells of *Shewanella alga* BrY were aerobically grown to the late exponential or early stationary phase. The cells were harvested by centrifugation, washed 3 times, and finally re-suspended in a phosphate buffered saline solution (PBS). The cells were starved by aseptically stirring on a magnetic stir plate at room temperature.

Fe-Coating of Quartz Sand. Quartz sand (40 mesh) was washed with 1 N HNO_3 and rinsed with deionized water. To 100 g of washed sand, 200 mL of a solution of amorphous Fe(III)-hydroxide was added. The amorphous iron solution was freshly prepared by dissolving $FeCl_3$ in deionized water (34.5 g/L) and slowly adjusting to pH 7 using NaOH. The Fe(III)-hydroxide-sand mixture was vigorously shaken and allowed to settle for several hours. The overlying water was then decanted and replaced with a 0.1 mM NaCl-solution and allowed to settle again overnight. The next morning, the Fe(III)-hydroxide-NaCl solution was discarded completely and replaced with freshly made Fe(III)-hydroxide solution. This process was repeated four times before the coated sand was allowed to air dry. The dried sand was sieved to break up large aggregates and remove fine material.

Batch Experiments. Vials containing 9 mL of anaerobic minimal medium (Caccavo et al., 1996) and 1 g of Fe(III)-coated sand were prepared. All anaerobic media and a lactate stock solution were boiled and cooled down under an oxygen free N_2 (80 vol %)-CO_2 (20 vol %) atmosphere, sealed gas tight and sterilized by autoclaving. Starved bacteria were injected into each vial (except for the bacteria free controls). The initial cell concentration was approximately 5 x 10^6 colony-forming units per mL (CFU/mL). All additions were made maintaining anoxic conditions using sterile syringes and needles; the vials were incubated statically in the dark. After a sufficient amount of Fe(II) had been produced, 0.1 mL of a K_2CrO_4 solution was injected into the vials to give a final concentration of approximately 5 mg/L as Cr(VI). Cr(VI) concentrations in the aqueous phase were monitored over time. K_2CrO_4 was also injected into autoclaved vials and control vials. At the end of the experiments, the vials were opened and analyzed for Cr(VI) and total chromium in the aqueous phase and on the sand.

Column Studies. Column studies utilizing PTFE-stainless steel assemblies were used to investigate the capability of starved BrY cells to reduce surface associated Fe(III) in a more field realistic setting. The columns were filled with approximately 7 g of Fe-coated sand and covered with aluminum foil to prevent photo-oxidation. Anaerobic minimal medium supplemented with 1 mM lactate was pumped through the columns once a day, ensuring the exchange of approximately 2 pore volumes in each column. Initially each column was inoculated with approximately 5 x 10^7 CFU of starved BrY cells. After Fe(II) had been produced in the columns, a K_2CrO_4 solution was continuously injected into the columns, and influent and effluent concentrations of Cr(VI) were monitored. At the end of the experiment, the columns were destructively sampled and the sand was analyzed for Cr(VI) and total Cr.

Analytical Techniques. The concentration of Fe(II) and total Fe were determined spectrophotometrically using ferrozine (Lovley and Phillips, 1987). The assay was slightly modified and utilized 2.5 N HCl or 0.25 NH_2OH-HCl in 2.25 N HCl for the extraction.

Cr(VI) and total chromium were determined using methods modified from Greenberg et al. (1992) (Cr(VI) aqueous, and total Cr in aqueous samples and from sand), and Vitale et al. (1994) (Cr(VI) from sand). Sym-diphenylcarbazide was utilized for the spectrophotometrical determination of chromium at 540 nm.

RESULTS AND DISCUSSION

Microbial Fe(III)-reduction. The resuscitation of starved *S. alga* BrY cells with artificially Fe(III)-coated sand is shown in Figure 1. After an initial lag phase, the

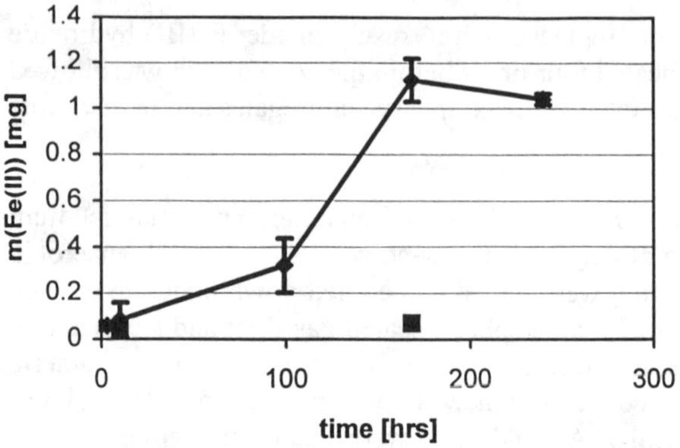

FIGURE 1. Resuscitation of starved *Shewanella alga* BrY cells on amorphous Fe(III)-oxyhydroxide coated sand. ◆ setups containing starved bacteria, ■ bacteria-free controls. The error bars represent the standard error of the mean.

injected cells of *S. alga* BrY reduced the surface associated Fe(III)-minerals, resulting in increasing Fe(II)-concentrations over time. The final mass of Fe(II) produced in batch - experiments (~1.1 mg Fe^{2+}) is equivalent to more than 80% of the total amount of Fe(III) added (~1.3 mg). Most of the Fe(II) produced remained associated with the sand surface as observed by Gorby et al. (1994). No increase in Fe(II)-concentrations was observed in bacteria-free controls or controls containing bacteria but no lactate, suggesting that the Fe(III)-reduction was due to microbial activity.

In columns inoculated with starved BrY cells and operated in semi-batch mode up to 80 % of the surface associated Fe(III) was reduced to Fe(II) after approximately 8 weeks and remained associated with the sand surface.

Cr(VI)-reduction by microbially reduced Fe(II). Batch vials containing microbially produced Fe(II), regardless of whether autoclaved (data not shown) or not, reduced the injected Cr(VI) rapidly while controls containing no or little Fe(II) (< 0.1 mg) did not show any detectable Cr(VI)-reduction (Figure 2). Even at the first sampling event, 30 seconds after injection of the K_2CrO_4-solution, the Cr(VI) concentrations in vials containing microbially reduced Fe(II) had decreased rapidly and continued to decrease over the duration of the experiment. The con-

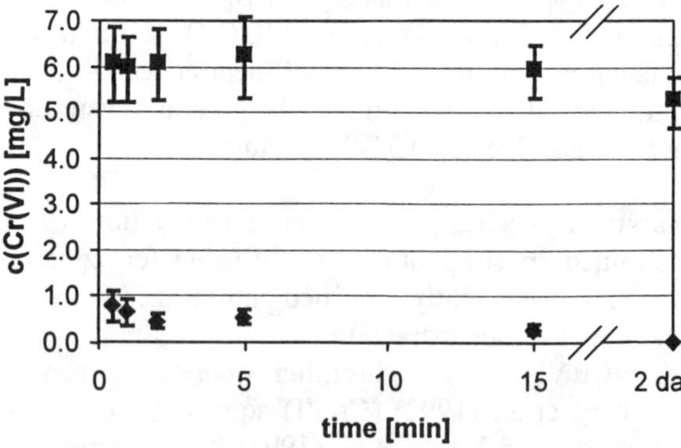

FIGURE 2. Aqueous phase Cr(VI) concentrations over time in vials containing microbially reduced Fe(II) (◆, n=6) and controls with no or little microbially reduced Fe(II) (< 0.1 mg) (■, n=2). The error bars represent the standard error of the mean.

trols, containing no or little Fe(II), showed relatively constant Cr(VI)-levels throughout the experiment (up to 2 days). This indicates that the reduction of Cr(VI) is due to the presence of microbially reduced Fe(II).

The columns containing microbially reduced Fe(II) showed a noticeable delay in the breakthrough of injected Cr(VI) as compared to a non-retarded tracer (Figure 3). The amount of Cr(VI) reduced was calculated as the difference between the area of the breakthrough curve for Cr(VI) and the integral over the Cr(VI) injection-profile, divided by the flow rate. The amount of Cr(VI) reduced in the column was calculated to be approximately 35 % of the total capacity for Cr(VI)-reduction (assuming a 3:1 ratio for Fe(II):Cr(VI), Blowes et al., 1997). This difference may be due to blocking of potential reaction sites by Fe-Cr-precipitates or the reaction of reduced Fe(II) with other constituents of the injection fluid (e.g. residual oxygen). More than 95 % of the chromium recovered was recovered as Cr(III) from the column sand (data not shown).

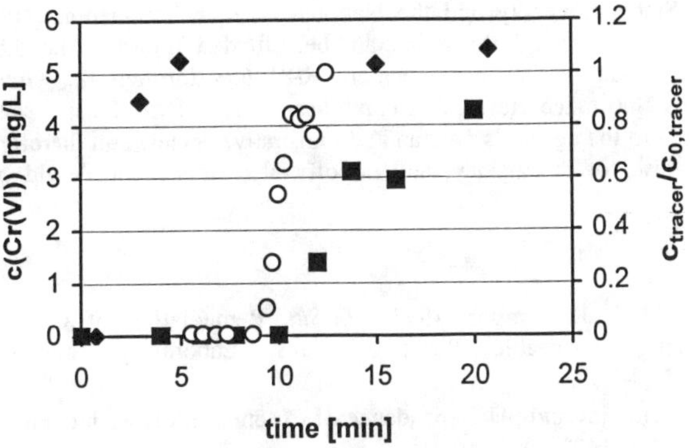

FIGURE 3. Cr(VI) concentrations in influent (◆) and effluent (■) of column containing microbially reduced Fe(II). Tracer breakthrough data (O) correspond to right y-axis.

CONCLUSIONS

Starved cells of *Shewanella alga* BrY can be resuscitated with artificially Fe(III)-coated sand as the electron acceptor, resulting in the production of surface associated Fe(II). The microbially produced Fe(II) almost instantaneously reduces the common groundwater contaminant Cr(VI) to Cr(III), which precipitates on the sand. The structure and stability of the precipitated chromium has been assessed by many authors (Blowes et al., 1997; Peterson et al., 1997). The precipitates have been described as $Cr(III)_xFe(III)_{1-x}$-(oxy)hydroxides. The Cr(III) ions appear to be stable under most conditions encountered in natural groundwater systems. These results suggest that metal-reducing bacteria can be used to reduce indigenous metals in or downstream of a contaminated aquifer to create a zone of redox-reactive Fe(II), which can in turn chemically reduce common groundwater contaminants such as heavy metals, radionuclides and chlorinated aliphatics. To further develop this technology, studies will be conducted that address the re-activation of depleted reactive iron zones. Column studies will be utilized to investigate whether this is possible and how long it will take to re-activate such zones by stimulating microbial Fe(III)-reduction.

The current results suggest that a permeable redox-reactive biobarrier technology based on the bioaugmentation with DMRB is possible.

ACKNOWLEDGMENTS

Thanks to Jennifer Nyman and Steven Hasslinger for help in the laboratory. This research is supported by the U.S. EPA Great Plains/Rocky Mountain Hazardous Substance Research Center headquartered at Kansas State University and the National Science Foundation (NSF Cooperative Agreement EEC-8907039). Although this article had been funded in part by the U.S. Environmental Protection Agency under assistance agreement R-819653 through the Great Plains/Rocky Mountain Hazardous Substance Research Center headquartered at Kansas State University, it has not been subjected to the agency's peer and administrative review and therefore may not necessarily reflect the views of the agency, and no official endorsement should be inferred.

REFERENCES

Blowes, D. W., C. J. Ptacek, and J. L. Jambor. 1997. "*In-Situ* Remediation of Cr(VI)-Contaminated Groundwater Using Permeable Reactive Walls: Laboratory Studies." *Environ.Sci.Technol. 31*: 3348-3357.

Buerge, I. J. and S. J. Hug. 1997. "Kinetics and pH Dependence of Chromium(VI) Reduction by Iron(II)." *Environ.Sci.Technol. 31*: 1426-1432.

Caccavo Jr., F. and G. J. Olson. 1995. "Cr(VI) Reduction and Remediation by the Dissimilatory Metal-Reducing Bacterium *Shewanella alga* strain BrY" In *Abstr. Emerging Technologies in Hazardous Waste Management*. VII. Division of Industrial and Engineering Chemistry, American Chemical Society, Washington D.C.

Caccavo Jr., F., N. Birger Ramsing and J.W. Costerton. 1996. "Morphological and Metabolic Responses to Starvation by the Dissimilatory Metal-Reducing Bacterium *Shewanella alga* BrY." *Appl.Environ.Microbiol. 62*: 4678-4682.

Cantrell, K. J, D. I. Kaplan and T. W. Wietsma. 1995. "Zero-Valent Iron for the In Situ Remediation of Selected Metals in Groundwater." *J.Haz.Materials 42*: 201-212.

Fortner, B. 1995. "Wall of Iron Purifies Contaminated Groundwater." *Water Environment and Technology*. 7: 14-17.

Gorby, Y. A., S. E. Amonette, and J.S. Fruchter. 1994. "Remediation of Contaminated Subsurface Materials by a Metal-Reducing Bacterium." In *Proceedings of the 33rd Hanford Symposium on Health and the Environment*, pp. 233-247. Battelle Press, Columbus, OH.

Greenberg, A. E., L.S. Clesceri, and A.D. Eaton (Eds.). 1992. *Standard Methods for the Examination of Water and Wastewater*. 18th ed., pp. 3-58-3-60, American Public Health Association, Washington, D.C.

Lovley, D. R. and E. J. P. Phillips. 1987. "Rapid Assay for Microbially Reducible Ferric Iron in Aquatic Sediments." *Appl.Environ.Microbiol. 53*: 1536-1540.

Peterson, M. L., A. F. White, G. E. Brown, Jr. and G. A. Parks. 1997. "Surface Passivation of Magnetite by Reaction with Aqueous Cr(VI): XAFS and TEM Results." *Environ.Sci.Technol. 31*: 1573-1576.

Scott, M. J., F. B. Metting, J. S. Fruchter and R. E. Wildung. 1998. "Research Investments Pays Off." *Soil Sci.Soc.Am.J. October 1998*: 6-13.

Sharp, R.R., R. Gerlach, and A. Cunningham. 1999. "Bacterial Transport Issues Related to Subsurface Biobarriers." In *In Situ and On-Site Bioremediation*. Proceedings of the Fifth International Symposium on In Situ and On-Site Bioremediation. April 19-22, 1999. San Diego, CA. Battelle Press, OH. In Press.

Vitale, R. J., G. R. Mussoline, J. C. Petura and B.R. James. 1994. "Hexavalent Chromium Extraction from Soils: Evaluation of an Alkaline Digestion Method." *Journal of Environmental Quality 23*: 1249-1256.

REMEDIATION OF A HEAVY METAL PLUME USING A REACTIVE WALL

Rick McGregor (Conor Pacific-WTI, Burlington, Canada)
Dave Blowes (University of Waterloo, Waterloo, Canada)
Ralph Ludwig (U.S. EPA, Ada, Oklahoma, USA)
Eric Pringle (Conor Pacific-WTI, Vancouver, Canada)
Martin Pomeroy (Environment Canada, Vancouver, Canada)

ABSTRACT: Historical storage of sulfide mineral ore concentrate at an industrial site in Western Canada has resulted in widespread contamination of underlying soil and groundwater. The oxidation of sulfide minerals has released significant quantities of heavy metals including Cd, Cu, Ni, Pb, and Zn into the groundwater. The impacted groundwater currently threatens a nearby marine ecosystem. In March 1997, a pilot-scale compost-based sulfate-reducing porous reactive wall was installed within a dissolved heavy metal plume at the site. The porous reactive wall uses sulfate-reducing bacteria to convert dissolved sulfate in the heavy metal plume to sulfide. As a consequence, heavy metals are precipitated out as insoluble metal sulfides. During the trenching phase of construction, biodegradable (guar gum) slurry was used to stabilize trench walls to allow for effective emplacement of the reactive media. The guar gum was observed to disappear from within the wall over a 3-4 week period following installation. Monitoring of the wall to date has shown significant decreases in heavy metal concentrations within and down-gradient of the wall suggesting that sulfate reduction and metal sulfide precipitation are effectively occurring. Dissolved copper concentrations have been observed to decrease from approximately 2200 μg/L to less than 5 μg/L following treatment within the wall. Cd, Ni, Pb, and Zn concentrations have shown similar decreases with treatment concentrations generally being observed below instrument detection limits.

INTRODUCTION

Soils and groundwater at an industrial site in Greater Vancouver, Canada are extensively contaminated with heavy metals including cadmium, copper, nickel, lead and zinc. The heavy metal contamination has resulted from the oxidation of sulfide mineral ore concentrate historically stored on site. Heavy metals and acidity in the groundwater currently pose a potential threat to a nearby marine ecosystem located immediately down-gradient of the site.

A permeable reactive pilot test wall was installed within the groundwater plume in late March 1997 to evaluate the potential application of reactive wall technology at the site. Permeable reactive treatment wall technology is an emerging technology which is being increasingly employed for treatment of contaminated groundwater (Blowes and Ptacek, 1992; Gillham and O'Hannesin, 1992; Robertson and Cherry, 1995; Powell et al., 1995; Baker et al., 1996; USEPA, 1998). The technology involves the placement of a permeable wall containing reactive materials in the path of the contaminant plume such that target contaminants are removed or transformed while groundwater flows through the wall unimpeded.

METHODS

Selection of Reactive Material. The reactive media used in the test wall at the Site was a compost-based media designed to microbially reduce sulfates to sulfides and thereby precipitate out target heavy metals. The media used for the permeable reactive wall needed to satisfy several criteria. Firstly, the media needed to be sufficiently permeable to accommodate the groundwater flux at the site yet be sufficiently reactive to effectively precipitate out the heavy metals. The media would also need to sustain the desired permeability and reactivity properties for a sufficient period of time in order to be cost-effective relative to alternative treatment technologies.

The selection of the organic carbon (compost) source of material for the field study was based on laboratory experiments conducted by Waybrant et al. (1998) and on the results of another reactive wall installed in Sudbury Canada (Benner et al., 1998). The results of these studies indicated that leaf compost could be an effective and long-term source of organic carbon as well as nitrogen, phosphate and other trace elements for the sulfate-reducing bacteria.

Studies of the aquifer material at the site indicated that the hydraulic conductivity of the reactive wall needed to be greater than 0.03 cm/sec. Laboratory hydraulic conductivity tests indicated a mixture containing approximately 85% pea gravel was required to obtain a hydraulic conductivity of 0.15 cm/sec. To achieve a relatively homogeneous mixture in the field, the compost and pea gravel were thoroughly mixed using a backhoe prior to emplacement within the wall.

Wall Installation. Soils at the site are comprised primarily of sandy gravel with minor fines and cobbles. Groundwater flow at the site is tidally influenced with the net groundwater flow being in a southerly direction toward a marine inlet. The test reactive wall was installed in an area of high groundwater and soil contamination using cut and fill excavation methods. As the soil was removed from the trench, the reactive mixture was placed in the trench using a excavator bucket. Due to unstable soil conditions encountered during trenching, a biodegradable slurry was used to keep the trench open during emplacement of the mixture. The biodegradable slurry used was a low viscosity Guar Gum mixture which, was pumped into the excavation during trenching. The test reactive wall was installed to a depth of approximately 6.7 metres below grade with a length of 10 metres and a thickness of 2.5 metres.

Sampling of Monitoring Wells. A total of seventeen (17) bundle-style multi-level piezometers were installed up-gradient, within, and down-gradient of the reactive wall to monitor for various chemical and physical parameters (Figure 1). Each bundle was comprised of seven (7) sampling points spaced at approximately

0.5 metre intervals below the water table. The deepest sampling point in each bundle was located below the reactive wall. Three sets of five bundles were installed perpendicular to the reactive wall. A separate bundle was installed on each side of the reactive wall to monitor for possible "by-passing" of the reactive wall (Figure 1)

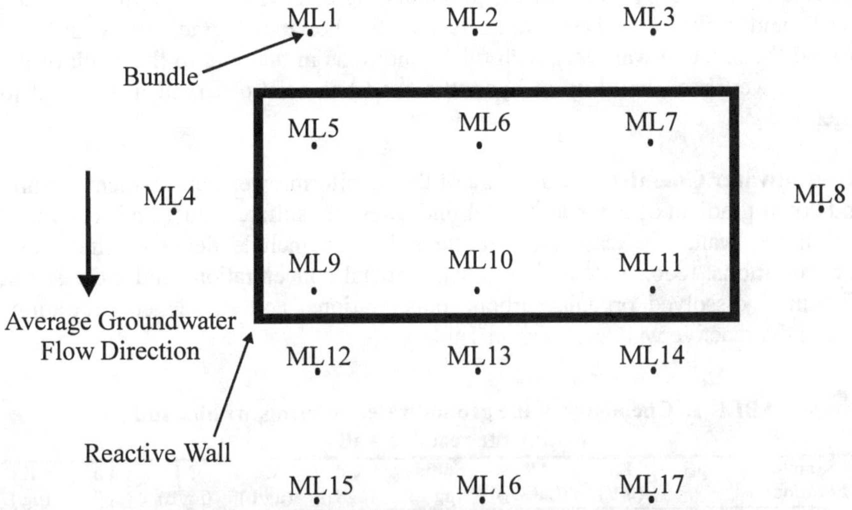

FIGURE 1. Location of monitoring bundles relative to the reactive wall in plan view

Samples of groundwater were collected every three months beginning one month after installation of the reactive wall. Water samples were collected using a peristaltic pump and were filtered in the field using 0.45 µm filters. Samples to be analyzed for cations were acidified to pH<1.5 using nitric acid. Samples to be analyzed for dissolved organic carbon (DOC) were acidified using 6 N sulfuric acid. Samples to be analyzed for anions were left unpreserved. All samples were refrigerated at the field site and stored refrigerated until analyzed. Laboratory spiked samples, blanks, and replicate samples were submitted with each batch of samples in addition to internal laboratory spikes and blanks. Charge-balance calculations were also conducted on all samples. All samples were determined to have a charge-balance difference of less than 8%.

Determinations of pH (Cole Parmer 55500-14 gel filled combination electrode) and E_H (ORION 9678BN platinum electrode) were made in the field using sealed flow-through cells. The pH electrode was calibrated using pH 4.0 and 7.0 buffer solutions (traceable to NIST). The E_H electrode was checked using Zobell's solution (Zobell, 1946; Nordstrom et al. 1977). Determinations of alkalinity were made in the field by titration with standardized H_2SO_4 using a digital titrator. Sulfide measurements were also made in the field using a portable spectrophotometer (HACH DR2010™).

RESULTS AND DISCUSSION

Hydraulic Properties. Hydraulic conductivity tests completed on piezometers within the treatment wall using the method of Hvorslev (1951) indicated that the average horizontal hydraulic conductivity was approximately 0.18 cm/sec. Measurements of the hydraulic heads within and outside of the reactive wall indicated that the overall vertical gradients were downwards varying between $5x10^{-5}$ and $1.5x10^{-4}$. These gradients like the horizontal gradients within and around the reactive wall vary with tidal conditions in the inlet to the south of the site. However, no build-up of hydraulic head has been observed in the wall to date.

Groundwater Chemistry. Sampling of the monitoring wells up-gradient, within, and down-gradient of the reactive wall indicates that sulfate reduction is occurring within the wall. Indicators of sulfate reduction include decreases in sulfate concentrations, redox potential, and heavy metal concentrations and increases in alkalinity, dissolved organic carbon concentrations, and sulfide concentrations within the reactive wall as shown in Table 1.

TABLE 1. Chemistry of the groundwater entering, within, and exiting the reactive wall

Sample Location*	pH	E_H (mV)	Alk. ($CaCO_3$)	Sulfide (µg/L)	Cd (µg/L)	Cu (µg/L)	Ni (µg/L)	Pb (µg/L)	Zn (µg/L)
Up-gradient (ML-2)	6.36	452	99	<1.0	23	2250	155	4.4	3530
Within -front (ML-6)	6.43	26	293	1700	<0.1	2.6	2.0	4.3	<20
Within-back (ML-10)	6.50	6.5	344	610	<0.1	1.9	0.8	3.3	<20
Down-gradient (ML-13)	6.47	100	344	130	<0.1	3.4	1.3	3.0	<20

* Average values for monitoring points at that location.

The metals of concern at the site are cadmium [Cd], copper [Cu], nickel [Ni], lead [Pb], and zinc [Zn]. Dissolved concentrations within the groundwater vary spatially and therefore for the sake of simplicity, the average concentration values for all monitoring wells at each sampling location (i.e. up-gradient, within the wall, and down-gradient) are used for discussion. Treatment within the wall has reduced the average heavy metal concentrations for copper and zinc by over three orders of magnitude. The data used in the following discussions is based on a sampling event undertaken in June 1998 approximately 15 months following wall installation.

Concentrations of dissolved cadmium flowing into the reactive wall averaged 23 µg/L at piezometer bundle location ML-2 (Table 1). Cadmium concentrations within and immediately down-gradient of the wall were below the detection limit (<0.1 µg/L). Copper and nickel concentrations in the groundwater entering the wall at ML-2 averaged 2250 µg/L and 155 µg/L, respectively (Table 1). These concentrations decreased to averages of 1.9 µg/L and 0.8 µg/L, respectively within the wall and 3.4 μg/L and 1.3 μg/L, respectively immediately down-gradient of the wall. The increase in concentrations down-gradient of the wall is due to the effects of infiltration of precipitation through contaminated soil within the vadose zone and the subsequent diffusion of contaminants into the treated groundwater plume emerging from the wall.

Dissolved lead and zinc concentrations entering the wall (ML-2) averaged 4.4 μg/L and 3530 μg/L, respectively. Concentrations of these metals decreased to an average of 3.0 μg/L for lead and less than 20 μg/L (i.e. below the detection limit) for zinc within the wall.

The pH of the groundwater increases slightly as it passes through the reactive wall from an average of 6.36 to an average of 6.50 within the wall. Increases in alkalinity (as CaCO3) are also observed as the groundwater passes through the reactive wall. Alkalinity increases from an average of 99 mg/L to an average of 344 mg/L as the groundwater exits the reactive wall. This increase in alkalinity is due to complex reactions between the microbiological community and buffering reactions with the limestone within the reactive wall. The net effect of these reactions is an overall increase in pH-buffering capacity of the groundwater. Often associated with heavy metal contamination is high dissolved concentrations of Fe(II), which from sparingly soluble oxyhydroxides upon oxidation. These reactions result in the production of acid which, can be buffered by the alkalinity produced as the groundwater passes through the reactive wall.

CONCLUSIONS

Monitoring of geochemical and hydrogeological parameters 15 months after installation of the wall indicates that sulfate reduction and metal sulfide precipitation is taking place within the reactive wall. Metals of concern including cadmium, copper, nickel, lead and zinc are being effectively attenuated by reactions within the wall. Hydraulic conductivity and hydraulic pressure measurements indicate that the reactive mixture is not impeding or diverting groundwater flow but is instead acting as a zone of high hydraulic conductivity relative to the surrounding aquifer.

AKNOWLEDGEMENTS

Funding for this project was provided by Environment Canada and the Burlington Environment Technology Office (BETO). The authors would like to thank R. Booth and P. Kluckner for their editorial assistance as well as M. Choi, M. Baker and J. Bain for their assistance in sampling. Interpretation and modeling of the data was done with the assistance of S. Benner and C. Ptacek.

DISCLAIMER

The views expressed in this paper are those of the individual authors and do not necessarily reflect the views and policies of the U.S. Environmental Protection Agency (EPA) or Environment Canada.

REFERENCES

Baker, M.J., D.W. Blowes and C.J. Ptacek. 1996. "Development of a Reactive Mixture to Remove Phosphorous form On-Site Wastewater Disposal Systems." In *Proceedings Disposal Trenches, Pre-Treatment and Re-Use of Wastewater*, Waterloo, Ontario. 51-56.

Benner, S.G., D.W. Blowes and C.J. Ptacek. 1998. "A Full-Scale Porous Reactive Wall for Prevention of Acid Mine Drainage." *Ground Water Monitoring and Remediation*. 17(4): 99-107.

Blowes, D.W. and C.J. Ptacek. 1992. "Geochemical Remediation of Groundwater by Permeable Reactive Walls: Removal of Chromate by Reduction with Iron Bearing Solids." In *Proceedings Subsurface Restoration Conference, Third International Conference on Groundwater Quality Research*. Dallas Texas. 214-216.

Gillham, R.W. and S.F. O'Hannesin. 1992. "Metal-catalysed Abiotic Degradation of Halogenated Organic Compounds." In *Modern Trends in Hydrogeology*, Hamilton, Canada, IAH, 94-103.

Hvorslev, M.J. 1951. Time Lag and Soil Permeability in Groundwater Observations." *U.S. Army Corps Engineers. Waterways Experimental Station Bull. 36*, Vicksburg, Miss.

Nordstrom, D.K., E.A. Jenne and J.W. Ball. 1977. "Thermochemical redox equilibria of Zobell's solution." In *Chemical Modeling of Natural Systems*, eds. A.J. Everett and J.W. Ball, 51-79. Washington, D.C., American Chemical Society.

Powell, M.P., R.W. Puls, S.K. Hightower and D.A. Sabatini 1995. Coupled Iron Corrosion and Chromate Reduction: Mechanisms for Subsurface Remediation *Environ. Sci. Technol.* 29:1913-1922

Roberston, W.D. and J.A. Cherry, 1995. "In Situ Denitrification of Septic System Nitrate Using Reactive Porous Media Barriers: Field Trials." *Ground Water.* 31:99-111.

USEPA. 1998. Permeable Reactive Barrier Technologies for Contaminant Remediation *EPA/600/R-98/125*

Waybrant, K.R., D.W. Blowes and C.J. Ptacek. 1998. "Selection of Reactive mixtures for Use in Permeable Reactive Walls for Treatment of Mine Drainage." *Environ. Sci. Technol.* 32:1972-1979.

Zobell, C.E. 1946. "Studies of redox potential of marine sediments." *Bul. Amer. Ass. Petrol. Geol.* 30(4): 477-513.

ASSESSING METAL BIOAVAILABILITY IN SOILS USING LUMINESCENCE-BASED MICROBIAL BIOSENSORS

[1]**Preston, S.**, [1]Hall, J.M., [1]Rattray, E.A.S., [2]Chaudri, A.M., [2]McGrath, S.P., [1]Killham, K. and [1]Paton, G.I.
[1]University of Aberdeen, Aberdeen, AB24 3UU, U.K.
[2]Integrated Approach to Crop Research-Rothamsted, Harpenden, AL5 2JQ, U.K.

Abstract: Laboratory and field-scale experiments were carried out to investigate the bioavailable fraction of heavy metal pollutants using luminescence-based microbial biosensors. Laboratory experiments demonstrated that 90 % of the variation in the biosensor response was explained by the free metal ion concentration in a range of Zn and Cu aqueous standards at different pH values, the only exception being Cu^{2+} at pH 7.0. The free metal ion concentration was predicted using GEOCHEM-PC. Similarly, field-scale experiments where Zn was the main pollutant present showed that overall it was the concentration of soil solution Zn^{2+} which explained most of the variation in the response of the biosensors. The concentration of Zn^{2+} in soil solution was measured using a calcium saturated cation exchange resin. Luminescence-based microbial biosensors were shown to provide rapid, inexpensive techniques to predict the bioavailable fraction of heavy metals in environmental samples. Such biosensors may therefore serve as powerful tools in the assessment of toxicity and bioremediation potential of contaminated sites resulting in a more effective management of waste.

INTRODUCTION

Future levels of heavy metals in agricultural soils may increase as a result of an increase in the application of sewage sludge to land. Current guidelines concerning maximum permissible concentration of metals in soil are based on total metal concentrations. However, these guidelines have been questioned since recent evidence suggests that it is the free metal ion concentration which is toxic to organisms (McGrath, 1994).

Interest in the application of biosensors for environmental monitoring has increased due to their rapidity, low cost and portability compared to conventional analytical techniques. Biosensors are analytical devices consisting of a biological sensing element (e.g., microorganisms, enzymes etc.) either linked with or integrated within a physicochemical transducer (e.g., optical, electrochemical etc.). Biosensors have the added advantage of being able to operate under a range of environmental conditions (e.g., pH, exchangeable cations, organic matter etc.) that contribute to the bioavailability of pollutants.

Luminescence-based biosensors have been increasingly used as a rapid, sensitive and cost-effective technique in toxicity testing (Steinberg *et al.*, 1995). Microtox®, an international benchmark in environmental toxicity testing, is based on the naturally luminescent marine bacterium *Vibrio fischeri* (Azur Environmental Ltd., Berkshire). The assay measures bacterial luminescence, where decreases in luminescence reflect decreases in cellular metabolic activity due to the action of a pollutant. However, the assay is less applicable to non-marine environments, as it requires a neutral pH and the addition of a saline buffer which may alter the speciation of a pollutant. The ability to clone the genes responsible for luminescence into an organism of choice enables the construction of ecologically relevant biosensors. Such microbial biosensors have been developed for the soil environment and successfully applied in soil toxicity testing (Paton *et al.*, 1997).

The aim of this paper is to relate the biosensor response to the bioavailable fraction of metals in a simple aqueous system initially and then in the interstitial pore waters extracted from two long term sewage sludge field trials.

MATERIALS AND METHODS

Preparation of Aqueous Metal Solutions. Stock solutions (1111 mg/L) of Zn and Cu were prepared in double deionised water using sulphate salts of the respective metal, acidified with 1 ml/L 1 M HNO_3, and stored at 4°C. A range of Zn and Cu test solutions was prepared from the stock solutions in double deionised water and the pH value of the solutions adjusted to 5.5 ± 0.02 using HCl and NaOH. Free metal ion concentrations in the test solutions were predicted using GEOCHEM-PC.

Sampling of Experimental Field Sites. Two sewage sludge field trials were sampled: Braunschweig, Germany and Gleadthorpe, U.K. Both sites had received sewage sludge contaminated primarily with Zn and Cu. Soil solutions from replicate treatment plots were obtained using soil moisture samplers (Rhizosphere Research Products, Eijelkamp, Holland) as described by Knight *et al.* (1998).

Chemical Analysis of Field Sites. Measurements of soil solution pH were made using a glass electrode and of dissolved organic carbon (DOC) using an automated PPM LABTOC analyser. Metal concentrations were determined by Inductively-Coupled Plasma Atomic Emission Spectrometry (ICP-AES) and Graphite Furnace Atomic Absorption Spectrometry (GF-AAS). Free metal ion concentrations were determined using a cation exchange resin method (Holm *et al.*, 1995).

Determination of Metal Bioavailability Using Microbial Biosensors. The following luminescence-based microbial biosensors were used to determine metal bioavailability in aqueous solutions and soil solutions: *Escherichia coli* HB101 pUCD607 (Rattray *et al.*, 1990) and/or *Pseudomonas fluorescens* 10586 pUCD607 (Amin-Hanjani, 1993). The biosensors were stored as freeze-dried cultures and resuscitated prior to use according to standard laboratory protocols (Paton *et al.*, 1997). The assays consisted of pipetting 50 μl or 100 μl of resuscitated cells into cuvettes containing 450 or 900 μl of the test solution, respectively. Cells were added at 15 s intervals between samples and mixed with the test solutions during the initial 5 s. The luminescence of the samples was measured on a Bio-Orbit 1231 or 1251 luminometer (Labtech International, Uckfield, U.K.) after exposure times ranging from 15 to 145 minutes. All assays were carried out in triplicate.

Data Analysis. The toxicity of the test solutions was determined by expressing the luminescence response obtained as a percentage of the control luminescence. The control consisted of either double deionised water at pH 5.5, in the case of aqueous solutions, or of soil solutions from control plots, in the case of the experimental field sites. Statistical analyses were carried out using Microsoft Excel (version 5), SigmaPlot (version 3.06) and Genstat 5 for Windows (release 3).

RESULTS AND DISCUSSION

The toxicity of the Zn and Cu standards to the *E. coli* HB101 pUCD607 biosensor was found to increase as the pH of the solutions decreased (Figure 1).

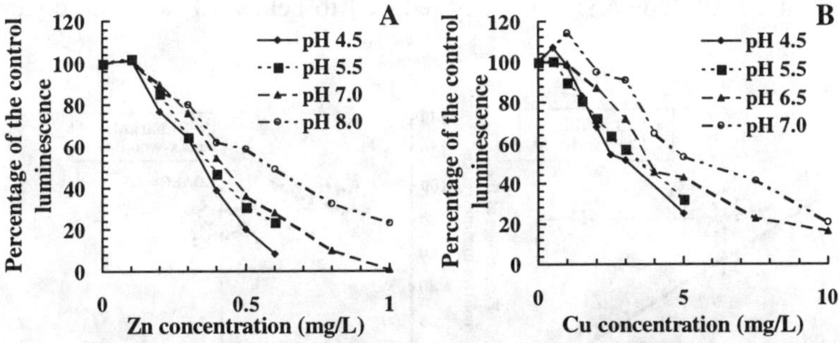

FIGURE 1. Biosensor response to Cu and Zn at different pH values.

The increased toxicity of Zn and Cu at lower pH values was related to an increase in the free metal ion concentration predicted by GEOCHEM-PC. The free metal ion concentration explained 90 % of the variation in the biosensor response (Figure 2), the only exception being Cu^{2+}, which at pH 7.0 was found to be less toxic than $CuOH^{+}$, the latter explaining 94 % of the biosensor response at pH 7.0.

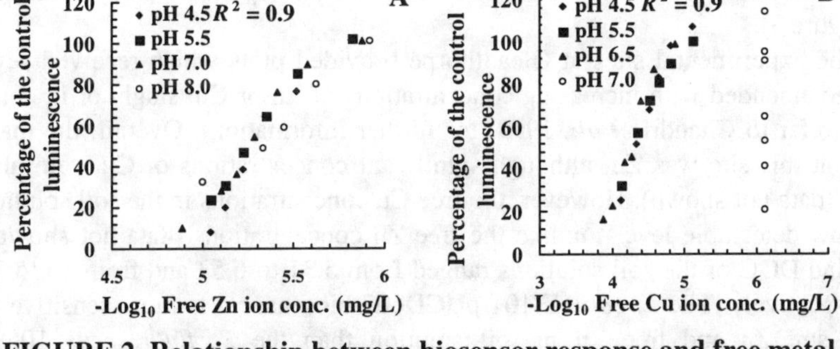

FIGURE 2. Relationship between biosensor response and free metal ion concentration (omitting Cu^{2+} data at pH 7.0 in the regression analysis).

The plots receiving sewage sludge at the field trial in Braunschweig, Germany, were found to be mainly polluted with Zn (refer to McGrath *et al.*, 1999 for further information). The dissolved organic carbon (DOC) in the soil solutions ranged from 21.5 to 49 mg/L. The experimental plots were situated on two sites with contrasting pH, an old arable site with a soil solution pH between 5.2 and 6.9 and an ex-woodland site with a soil solution pH between 4.4 and 5.1. This resulted in a higher concentration of free Zn ions in the soil solution in the ex-woodland site compared to the arable site despite the fact that both sites had comparable concentrations of total soil Zn (Figure 3). The importance of pH in increasing the concentrations of free metal ions has been shown previously (Adams *et al.*, 1984). The differences in free Zn ions between the sites accounted for the fact that the luminescence of the *P. fluorescens* 10586 pUCD607 biosensor was never significantly reduced in samples obtained from the old arable site (Figure 3). The ex-woodland site was quite different, the luminescence of the *P. fluorescens* 10586 pUCD607 being significantly reduced to below 10 % of the control (Figure 3).

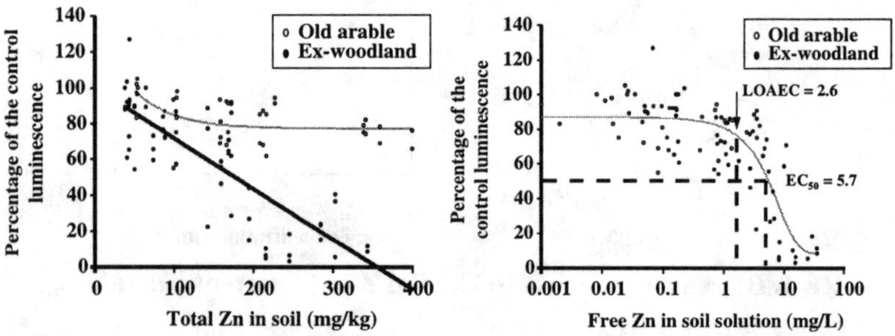

FIGURE 3. Relationship between biosensor response and either the total Zn in soil or the free Zn in soil solution.

Statistical analyses showed that total Zn in soil explained only 32 % of the variance in the luminescence of the *P. fluorescens* 10586 pUCD607 biosensor, while free Zn in soil solution explained more than 72 % of the variance in its luminescence. The latter resulted in the data being adequately described by a single curve (Figure 3).

The experimental sites at Gleadthorpe provided plots which received sewage sludge amended with increasing concentrations of Zn or Cu, singly or in combination (refer to Chaudri *et al.*, 1999, for further information). Overall, the main pollutant at this site was Zn, although significant concentrations of Cu were also observed (data not shown). However, the free Cu concentrations in the soil solution were below detectable levels, unlike the free Zn concentrations (data not shown). The pH and DOC of the soil solutions ranged from 5.31 to 6.53 and from 15 to 36 mg/L, respectively. The *E. coli* HB101 pUCD607 biosensor was more sensitive to the total soil Zn and free Zn in soil solution than the *P. fluorescens* 10586 pUCD607 biosensor (Figure 4). This is probably due to *E. coli* occupying a very different ecological niche to *P. fluorescens*. Non-linear regression analyses showed that the total soil Zn concentration explained 85 and 80 % of the variance

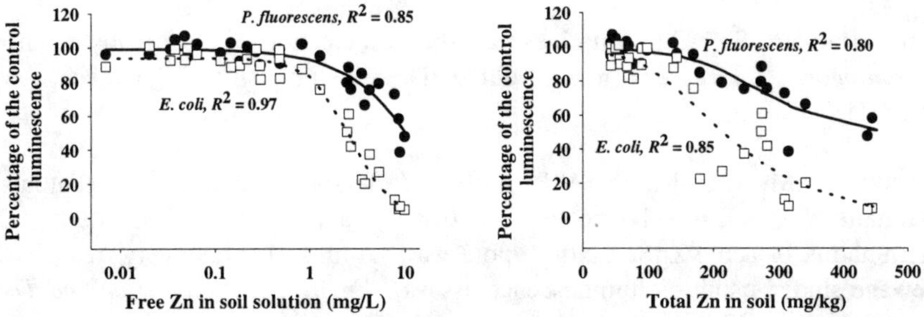

FIGURE 4. Relationship between the response of the biosensors and either the total Zn in soil or the free Zn in soil solution.

in the luminescence response of the *E. coli* HB101 pUCD607 biosensor and of the *P. fluorescens* 10586 pUCD607 biosensor, respectively (Figure 4). However, the concentration of free Zn in soil solution explained 97 and 85 % of the variance in the luminescence response of the *E. coli* HB101 biosensor and of the *P. fluorescens* 10586 biosensor, respectively (Figure 4). The concentration of free Zn in soil solution was therefore identified as the toxic species to the biosensors.

CONCLUSIONS

This study clearly shows that total metal concentrations in soil do not reflect the bioavailability of the metal pollutants. This raises serious issues about whether the current guidelines on metal levels in soils are sufficiently protective since they are based on total soil metal concentrations, with no consideration given to the free metal ion concentrations. The effective management of waste and toxicity assessment of contaminated sites with a view to reclamation can only succeed if a proper evaluation of toxicity is carried out. Any toxicity assessment must therefore examine the bioavailable fraction of pollutants using either conventional or novel techniques. In this study, the luminescence-based microbial biosensors were shown to provide a rapid, reliable and inexpensive technique to predict the bioavailable fraction of metal pollutants in soils. These characteristics make them powerful tools in the assessment and monitoring of toxicity in metal contaminated sites.

ACKNOWLEDGEMENTS

The authors would like to thank the Biotechnological and Biological Sciences Research Council and the DTI LINK Programme for funding this work.

REFERENCES

Adams, T.McM., S.P. McGrath, J.R. Sanders. 1984. "The effect of soil pH on solubilities and uptake into ryegrass of zinc, copper and nickel added to soils. In

T.D. Lekkas (Ed.), *Heavy Metals in the Environment*, Vol 1, pp. 484-486. Athens, C.E.P., Edinburgh.

Amin-Hanjani, S. 1992. "Luminescence-based detection of genetically modified *Pseudomonas fluorescens* in soil." Ph.D. Thesis. University of Aberdeen, Aberdeen, UK.

Chaudri, A.M., B.P. Knight, V.L. Barbosa-Jefferson, S. Preston, G.I. Paton, K. Killham, N. Coad, F.A. Nicholson, B.J. Chambers and S.P. McGrath. 1999. "Determination of acute Zn toxicity in pore water from soils previously treated with sewage sludge using bioluminescence assays." *Environmental Science and Technology* (Accepted).

Holm, P.E., T.H. Christensen, J.C. Tjell and S.P. McGrath. 1995. "Speciation of cadmium and zinc with applications to soil solutions." *Journal of Environmental Quality. 24*: 183-190.

Knight, B.P., A.M. Chaudri, S.P. McGrath and K.E. Giller. 1998. "Determination of chemical availability of cadmium and zinc using inert soil moisture samplers." *Environmental Pollution. 99*:293-298.

McGrath, S.P. 1994. "Effects of Heavy metals from sewage sludge on soil microbes in agricultural ecosystems." In S.M. Ross (Ed.), *Toxic Metals in Soil-Plant Systems*, pp. 247-274. John Wiley & Sons Ltd, New York.

McGrath, S.P., B. Knight, K. Killham, S. Preston, and G.I. Paton. 1999. "Assessment of the toxicity of metals in soils amended with sewage sludge using a chemical speciation technique and a *lux*-based biosensor." *Environmental Toxicology and Chemistry. 18*: 00-00.

Paton, G.I., E.A.S. Rattray, C.D. Campbell, M.S. Cresser, L.A. Glover, J.C.L. Meeussen and K. Killham. 1997. "Use of genetically modified microbial biosensors for soil ecotoxicity testing." In C. Pankhurst, B. Doube and V. Gupta (Eds.), *Biological Indicators of Soil Health and Sustainable Productivity*, pp. 397-418. CAB International.

Rattray, E.A.S., J.L. Prosser, K. Killham and L.A. Glover. 1990. "Luminescence-based nonextractive technique for *in situ* detection of *Escherichia coli* in soil." *Applied and Environmental Microbiology. 56*: 3368-3374.

Steinberg, S.M., E.J. Poziomek, W.H. Englemann and K.R. Rogers. 1995. "A review of environmental applications of bioluminescence measurements." *Chemosphere. 30*: 2155-2197.

MICROBIOLOGICAL TREATMENT OF CYANIDE, METALS, AND NITRATES IN MINING BYPRODUCTS

Suzzann Nordwick, Marietta Canty and Randy Hiebert,
(MSE Technology Applications, Inc., Butte, Montana)
Leslie Thompson, (Pintail Systems, Inc., Aurora, Colorado)

ABSTRACT: An innovative biological treatment system designed to detoxify cyanide and nitrate and immobilize metals from cyanide leach operations is described. A pilot-plant demonstration was conducted to detoxify cyanide and immobilize several metals from a process stream at the McCoy/Cove Mine. This technology demonstration was successful in detoxifying cyanide and effective in removing metals. The pH was consistently neutralized and nitrates were successfully removed. Results show that biological treatment offers an effective, innovative method for the treatment of mining byproducts.

INTRODUCTION

Typical contaminants in mine process waters include cyanide, metals, and nitrates. Cyanide is used in the mining industry to extract precious metals from ores. Cyanide also forms strong complexes with other metals, which can result in metal contamination. Nitrates arise from blasting agents or from reclamation materials such as fertilizer. Cyanide, metals, and nitrates can contribute to environmental problems. Conventional treatments tend to be expensive and chemical intensive. Consequently, the development of economical, effective treatment processes for mine byproduct remediation is a critical issue facing industry today. The conventional heap leach detoxification process is to rinse spent ore with fresh or treated water until compliance standards are achieved. This treatment process can require multiple pore volumes of water and may take several years.

An innovative technology has been developed to biologically detoxify cyanide and nitrates and remove metals from mining byproducts. This process is referred to as the biocyanide technology. It is capable of detoxifying cyanide and nitrate and immobilizing metals in wastewater from agitated cyanide leach operations, as well as heap leach operations. This paper describes a pilot-scale demonstration conducted to detoxify cyanide and immobilize metals from a mine process waste stream. The demonstration was performed jointly under the Mine Waste Technology Program (MWTP), which is funded by the U.S. Environmental Protection Agency (EPA) and is jointly administered by the U.S. Department of Energy (DOE) and the Superfund Innovative Technology Evaluation (SITE) Demonstration Program. MSE Technology Applications, Inc., (MSE) was the contractor responsible for conducting the project. Pintail Systems, Inc. was the technology provider, and Science Applications International Corporation (SAIC) was the quality assurance/quality control contractor for the project.

This innovative microbiological detoxification process offers several advantages over traditional detoxification processes. In addition to cyanide detoxification, the process also detoxifies nitrates and precipitates metals. When used to detoxify a spent heap leach pad, microbiological detoxification greatly shortens the length of time required. Microbiological treatment also has a lower comparative treatment cost than conventional treatments.

MATERIALS AND METHODS

Background. Some of the species that are known to oxidize cyanide include species of the genera Actinomyces, Alcaligenes, Arthrobacter, Bacillus, Micrococcus, Neisseria, Paracoccus, Pseudomonas, and Thiobacillus (Buchanan 1974). The reaction sequence in bacteria for cyanide oxidation and subsequent nitrate reduction is postulated to occur as summarized below.

$$Fe(CN)_6^{4-} + 29H_20 + 6.5O_2 \text{ ---> } 2CO_3^{2-} + 12NH_3 + 2Fe(OH)_3(s) + 16H \quad (1)$$

$$M_xCN_y + 2H_2O + 0.5O_2 \text{ ---> } M/bacteria + HCO_3 + NH_3 \quad (2)$$

$$NH_3 \text{ ----> } NH_2OH \text{ ----> } H_xNO_y \text{ ----> } NO_2 \text{ ----> } NO_3 \quad (3)$$

$$NO_3 \text{ ----> } NO_2 \text{ ----> } NO \text{ ----> } N_2O \text{ ----> } N_2 \quad (4)$$

M = metal

Microbiological remediation of cyanide is accomplished by introducing cyanide-metabolizing bacteria to a contamination source or by stimulating indigenous bacteria through nutrient addition. For this project, Pintail Systems collected water samples from the mine site to isolate indigenous organisms capable of effectively degrading cyanide. Laboratory bioaugmentation studies were performed. During this process, bacteria were subjected to increasing concentrations of cyanide to select the most capable organisms. The bacteria selected from these studies were used to perform bench-scale tests. The results from the bench-scale tests were used to design the pilot-scale system that was constructed at the mine.

The pilot-scale plant consisted of a proprietary biological culture of microorganisms, immobilized on a porous ceramic filtration media. The process contained three flow trains, two of which were biological treatment trains and one served as a control train. Each train consisted of an aerobic reactor followed by two anaerobic reactors and a small aerobic polishing reactor. The aerobic and anaerobic reactors were inoculated with aerobic and anaerobic microbiological cultures, respectively. The control train was not inoculated, but it was permitted to establish its own culture.

A slipstream of treatment solution was taken from the mining process. The aerobic reactors were operated in downflow configurations, and the anaerobic reactors were operated in upflow configurations. This resulted in the anaerobic reactors being operated as plug-flow systems, and the aerobic reactors being operated as trickling filters.

The biological treatment trains were alternately fed the cyanide solution to be treated. During a train's down-cycle, the reactors would be prepared for the next treatment cycle by circulating a nutrient/culture solution through the system. This was accomplished separately for the aerobic and the anaerobic reactors. The treatment bacteria were maintained as live cultures in a continuous culture system. The control train processed the same solution as the treatment trains. The data from the control reactor was compared with that of the on-line treatment train to provide a measure of performance.

Site Description. The McCoy/Cove Mine operated by Echo Bay Minerals was the demonstration site. This is a working gold and silver mine located 35 miles southwest of Battle Mountain in Lander County, Nevada. The mine is situated on the east slope of a north-south mountain range. Leach pads extend from the base of the foothills outward onto the flat lands.

Project Objectives. The primary objective of the project was to significantly reduce the weak acid dissociable (WAD) cyanide concentration in the mine process water. Specifically, to successfully reduce WAD cyanide to <1.0 milligrams per liter (mg/L) WAD cyanide concentration in the treated effluent with a 90% statistical confidence interval.

In addition to the primary project objective, secondary objectives provided additional background data on the operating characteristics and treatment capabilities of the technology. Secondary objectives included determining the effectiveness of the biocyanide process in reducing heavy metals concentrations and determining the percent reduction of WAD and total cyanides.

RESULTS AND DISCUSSION

Operation of the demonstration unit started June 11, 1997 and ended 77 days later on August 26, 1997. Several analytical parameters were monitored to generate enough data to appropriately evaluate the technology. Collected data included total cyanide, WAD cyanide, metals, nitrates, pH, and temperature. Samples were collected and analyzed according to the project-specific Quality Assurance Project Plan (SAIC 1997).

The chemistry of cyanide is very complex. Many cyanide compounds and their derivatives can be found in mine effluents and tailings. In general, cyanides can be classified as free cyanide, simple cyanide, and complex cyanide. Free Cyanides include cyanide (CN^-) and hydrogen cyanide (HCN). An examples of a simple cyanide compound would be $A(CN)_x$ where A is a metal. And, an example of a complex cyanide would be $A_yM(CN)_x$ where A is an alkaline metal and M is a heavy metal.

The relative stability of cyanide complexes ranges from weak to strong. Total cyanide analysis includes simple and complex cyanide species. WAD cyanide analysis includes the simple and weak metal-cyano complexes but not the stronger complexes (Smith 1991).

WAD Cyanide. Distillation WAD cyanide was substantially removed during the field demonstration as shown in Table 1. The average influent concentration was 242 mg/L while the average effluent concentration was 44 mg/L. The percent removal of WAD cyanide through the biocyanide pilot-scale unit ranged from 45% to nearly 100%. The average observed removal efficiency over the entire test period was 82%. The system operated more optimally during the latter portion of the test period form July 18 to August 26, 1997. During this time, an average removal efficiency of 88.8% was observed. The average removal efficiency for the control reactor during the latter test period was less than zero indicating that WAD cyanide was not removed.

TABLE 1. Cyanide results.

	TOTAL CYANIDE	WAD CYANIDE
Entire Test Period		
Average influent concentration	284 mg/L	242 mg/L
Average control concentration	287 mg/L	251 mg/L
Average effluent concentration	57 mg/L	44 mg/L
90% success with 90% confidence limit	103 mg/L	78 mg/L
Average removal (effluent)	80.5%	81.6%
Average removal (control)	0.8%	8.1%
Latter Test Period		
Average influent concentration	258 mg/L	217 mg/L
Average control concentration	276 mg/L	225 mg/L
Average effluent concentration	30.1 mg/L	25.5 mg/L
90% success with 90% confidence limit	55.1 mg/L	46.2 mg/L
Average removal (effluent)	88.7%	88.8%
Average removal (control)	<0%	<0%

Total Cyanide. Total cyanide was measured with the same frequency as WAD cyanide. Total cyanide was substantially removed as shown in Table 1. The average influent concentration of total cyanide was 284 mg/L while the average effluent concentration was 57 mg/L. The percent removal of total cyanide ranged from 45% to near 100%. The average removal efficiency over the entire test period was 81%. The system operated more optimally during the latter portion of the test period form July 18 to August 26, 1997. During this time, an average removal efficiency of 88.7% was observed. The average removal efficiency for the control reactor during the latter test period was less than zero indicating that WAD cyanide was not removed.

Metals. During the field demonstration, arsenic, cadmium, copper, mercury, nickel, selenium, silver, and zinc were analyzed as shown in Table 2.

TABLE 2. Metals Results.

	Average Influent (ug/L)	Average Effluent (ug/L)	Average Effluent Removal
Arsenic	332	46.8	85.2%
Cadmium	51	2	94.1%
Mercury	179	5.1	97.0%
Selenium	255	84.9	67.4%
Silver	882	15.4	97.0%
Zinc	29	0.06	96.5%
	(mg/L)	**(mg/L)**	
Copper	150	53	66.4%
Nickel	1.7	0.8	55.3%

Metals concentrations in the effluent samples were substantially lower than the concentrations in the influent samples. Metal removal occurred through biomineralization, a process by which microorganisms mediate and catalyze inorganic reactions to form new mineral assemblages. Biomineralization is a surface process associated with microorganism cell walls and excreted cell products where the mineralization occurs. A biomineral is defined to include both biologically formed authigenic minerals and complex biostabilized materials that are often clay- or gel-like silicates, sulfide, and oxides of variable composition that are presumed to be thermodynamically metastable phases (Thompson 1996).

Metal removal from the treated water occurred when microorganisms within the biological reactors formed new mineral assemblages, which reduced the mobility of metals. In addition to biomineralization, sulfate-reducing bacteria within the biological reactors are credited with forming insoluble metal sulfide precipitates through the production of hydrogen sulfide.

The average removal efficiencies for cadmium, mercury, silver, and zinc were very high at 94.1%, 97%, 97%, and 96.5%, respectively. These removal efficiencies were strikingly higher than those of the control reactor, which were 2.2% for cadmium, 13.4% for mercury and less than zero for silver and zinc.

For arsenic, the average removal efficiency was 84.9% in the treatment reactors while the control reactor had an average removal efficiency of 4.9%.

Treatment for selenium, copper, and nickel resulted in removal efficiencies of 67.4%, 66.4%, and 55.3%, respectively. The comparative control removal efficiencies were 4.7% for selenium and less than zero for copper and nickel.

pH. The pH was measured in the field using a hand-held pH analyzer. The pH of the effluent samples was consistently a little above neutral, at about 7.9. Influent mine water tended to be basic, grater than 9, at an average pH of 9.7. The average pH in the control reactor effluent was 9.6. This was consistent with the influent, indicating the control reactor did not alter the pH of the process water. The biological process was very successful in lowering the pH to neutral values.

Nitrates. Nitrate removal was not a stated objective of the demonstration. However, a significant reduction in concentration was observed when nitrate was measured for one week during the middle of the test period. The average influent nitrate concentration was 30.3 mg/L while the average effluent concentration was 1.6 mg/L. The average concentration from the control was 30.6 mg/L.

CONCLUSIONS

Substantial reductions of WAD and total cyanide were observed over the 11-week test period. The influent WAD cyanide concentration was 242 mg/L, and the effective effluent WAD cyanide concentration was 78 mg/L This is a substantial reduction of WAD cyanide concentration proving the microbiological treatment process very successful.

For metal treatment, cadmium, mercury, silver, and zinc had the highest removal efficiencies followed by arsenic, selenium, copper, nickel, and cobalt. Differences in removal efficiencies were attributed to the metal-specific effectiveness of both the biomineralization process and sulfate-reducing bacteria activity. The microbiological treatment process was considered to be very effective in removing metals.

The pH was consistently neutralized, and nitrates were successfully removed from the mine process water.

Conducting this technology demonstration at operating precious metals mine proved that the biocyanide process offers an effective, innovative method for the treatment of mining byproducts.

REFERENCES

Buchanan, R.E. and Gibbons, N.E., *Bergey's Manual of Determinative Bacteriology*, 8thed, Williams and Wilkins Company, Baltimore, Maryland, 1974.

SAIC, *Quality Assurance Project Plan: MWTP Biocyanide Field Demonstration at the Echo Bay McCoy/Cove Mine*, July 1, 1997.

Smith, A., and Mudder, T., *Chemistry and Treatment of Cyanidation Wastes*, Mining Journal Books, Ltd., London, England, 1991.

Thompson, L.C., *Cyanide Biotreatment and Metal Biomineralization in Spent Ore Process Solutions at Inactive and Abandoned Mine Sites*, EPA Seminar Publications no. EPA/625/R-95/007, pp 30-53, October 1996.

STUDIES ON BIORECOVERY OF METALS FROM ACID MINE DRAINAGE

Rakesh Govind, Wei Yang (University of Cincinnati, Cincinnati, OH)
Henry H. Tabak (U.S. Environmental Protection Agency, Cincinnati, OH)

ABSTRACT: Acid mine drainage (AMD) is a common problem for the mining and smelting industries throughout the world. Typically AMD contains several dissolved heavy metals at high concentrations (> 100 ppm), more than 3g/L sulfate and pH of less than 3.0. High acidity and presence of several heavy metals make AMD a major environmental concern because of the possible deleterious effects of the drainage on vegetation and natural habitats. There are over 40,000 remote abandoned mines in the state of Montana alone, and several thousand such mines in other states of Pennsylvania, Ohio and West Virginia. Acid mine water, upwelling from these remote mines, mainly during the spring season, results in massive destruction of surrounding vegetation. Natural sources of acid mine water (AMD) can be subdivided into two main categories: (1) large basins, such as the Berkeley Pit in Butte, Montana, which contains 20 billion gallons; and (2) small remote mines, which overflow mainly during the spring season, and the flow rate of water in the spring is typically less than 100 gallons per minute. The overall strategy is to attempt recovery and recycle of the valuable metals from AMD present in large basins, and treat the acid water from remote mines using low-cost *in-situ* methods. In this paper, experimental results from two studies will be presented: (1) biological reduction of sulfate, present in acid mine water using active SRBs, and sequential precipitation of the metals as sulfides or hydroxides; and (2) *in-situ* precipitation of metal sulfides in remote mines by direct injection of organic, such as acetate.

INTRODUCTION

Acid mine water contains several metal sulfates at concentrations exceeding toxicity limits for plants, animals and humans. Typically, metals present in substantial concentrations are copper, zinc, manganese, aluminum, and iron. Conventional AMD treatments use lime to precipitate metals. These treatments present some serious limitations in terms of application and effectiveness. Bulk precipitation results in production of a mixture of unstable metal hydroxides and lead to increased disposal expense. In recent years, the use of sulfate reducing bacteria (SRB) to reduce sulfate to sulfide and precipitate metals in acid mine drainage has been proposed as an alternative to hydroxide precipitation (Govind *et al.* 1997).

$$2HS^- + M^{++} \longrightarrow MS_2 \text{ (sulfide precipitate)}$$

In this paper, biokinetics of SRBs has been studied at four pH levels of 5,6,7,8 and temperatures of 10°C, 25°C, 35°C and 45°C. Batch experiments were conducted at a constant pH and temperature, using acetate as the organic source, and concentrations of acetate, sulfate and biomass were measured periodically. The Monod equation was used to define the bioconversion of sulfate to sulfide, and the Monod parameters were determined as a function of pH and temperature. Three applications of this data will be discussed in this paper: (1) biorecovery of valuable metals, such as copper, zinc, and manganese from large bodies of acid mine water, such as the Berkeley Pit in Butte, Montana; (2) *in-situ* treatment of acid mine water in small remote mines by direct injection of volatile fatty acids; and (3) treatment of acid mine drainage in constructed wetlands.

BACKGROUND

Kinetic studies on microbial sulfate reduction of SRB with special carbon sources have been reported in many articles. Studies on kinetic parameters for certain SRB species growing on lactate as sole carbon and energy source were provided by Okabe and Characklis (1992), Reis *et. al.* (1992), Traore *et. al.* (1981 and 1982) and Ingvorsen *et. al.* (1984). Rate and stoichiometric parameters for propionate and butyrate-utilizing SRB were reported by several research groups (Widdel and Pfenning, 1982 and Schauder *et. al.*, 1986). Okabe and Characklis (1992) summarized a few reported rate and stoichiometric parameters for acetate-utilizing SRB.

Sulfate-reducing bacteria growth is sentitive to pH and temperature. Experiments at different controlled pH in which the hydrogen sulfide produced was continously stripped from the medium were carried out (Reis *et. al.*, 1992) to study the influence of pH on growth yield coefficient and on specific growth rate. It was found that growth yield coefficient increases with the pH, reaching an optimum value, 14.30g cell/ mol SO_4, at pH 6.6. The evolution of specific growth rate with the pH has the same pattern as that observed for the growth yield coefficient, that is, μ_{max} increases with pH, the maximum value being obtained at pH 6.6.

MATERIALS AND METHODS

Master Culture Reactor: A 5 liter anaerobic master culture reactor was set up in a constant temperature incubator to provide active sulfate-reducing bacteria (SRB) at high concentrations for all the experiments. The anaerobic mixed SRB culture obtained from municipal sewage district in Cincinnati, OH, was acclimated to sulfate and acetate in the master culture at 35°C and pH from 7-8. A draw and fill schedule was used to obtain a hydraulic retention time of 15 days. When conducting batch experiments, specific amount of culture was transferred to batch reactors under N_2 atmosphere from master culture. The nutrient medium was purged with 99.999% pure nitrogen for 2 hours before feeding to reactor, in order to exclude dissolved O_2 from the nutrient. The master culture reactor was continuously stirred and the gas produced in the reactor was bubbled through zinc acetate solution. The white precipitate of zinc sulfide in the zinc acetate solution was used to indicate the on-set of bacterial growth.

Batch Bioreactors: Four identical 1.5-liter batch reactors were designed and operated in this study. The reactors were water jacketed and temperature was maintained by recirculating water from a constant temperature water bath (Model 9005, Fisher Scientific, Inc.); pH was maintained at a desired set-point by automatic pH control with addition of acid solution (1*N* HCl). The contents of the reactors were kept completely mixed by a magnetic stir bar and the gas produced in the reactor was bubbled through zinc acetate solution. Liquid samples for the purpose of sulfate, acetate and cell analysis were drawn through a septum, located at the bottom of the reactors, by using a syringe fitted with a 0.2 μm filter to retain the suspended biomass.

Mine Reactor: An upflow anaerobic mine reactor, with an internal diameter of 15 cm and height 60 cm, was operated with acid mine water flowing upwards (hydraulic retention time of 39 days), to simulate the upwelling of mine water in remote mines. Acetate solution was injected near the bottom, and precipitates were collected at the bottom of the mine reactor. Liquid samples were withdrawn from the side of the reactor, along the height, and concentrations of sulfate, acetate and metals as well as pH was determined by analysis. The mine reactor was jacketed, and was operated at two temperatures: 10°C and 25°C.

Metal Precipitation Tests: Batch metal precipitation tests were conducted for both metal hydroxides and sulfides and simulation studies were conducted using the OLI computer program (OLI Systems, NJ). The pH range in which precipitation occurs at 10°C and 25°C were determined for metal hydroxides and sulfides, and calculated using the OLI program.

Analytical Methods: pH was measured using a pH controller (PHCN-37, Omega Engineering, Inc.) with a combination probe (PHE-6510, Omega Engineering, Inc.) and kept constant by using the pH controller in conjunction with a peristaltic pump (Model 203, Scientific Industries, Inc.). The pump was used to add acid solution (1*N* HCl). The pH probe was calibrated using buffer 7 and buffer 4 or 10 (Fisher Scientific, Pittsburgh, Pennsylvania).

The sulfate and acetate concentration in the reactors were determined by suppressor-type ion chromatography (Alltech Universal IC systems 490200) with conductivity detector. The column used for the analysis was Shondex I-524A(JM Science, Inc.) and the mobile solution was carbonate (0.03g/L) / bicarbonate(0.02g/L) at a flow rate of 1.7ml/min. The system was maintained at 35°C. The mobile solution was degassed by using 99.99% pure nitrogen and filtered before use through a vacuum filter system with 0.22μm membrane (MSI Micron Separations Inc.) to protect the column.

Concentration of metals was analyzed using ICAP 61E plasma emission spectrometer (Thermo Jarrell Ash Corporation). The emission source for this spectrometer is an inductively coupled argon plasma (ICAP) which is powered by a 2kW crytal controlled radio frequency (RF) generator operating at 27.12 MHz. The instrument was operated at high argon coolant flow (14 L/min) and radio frequency power of 1150 Watts. The auxiliary argon flow rate was 0.5 L/min. The sample delivering speed of the peristaltic pump was set at 1 mL/min, with a nebulizer argon pressure of 27 psi. Diffraction grating was set at 1510 grooves/mm. The wavelengths of emission light were: Fe (259.940 nm), Cu

(324.754 nm), Zn (213.856 nm), Al (308.215 nm), and Mn (257.610 nm). Liquid samples were filtered through a 0.22 μm filter to remove solids and diluted with 2% nitric acid solution. Six standard solutions of different concentrations from 0 ppm to 50 ppm were used to obtain linear calibration curves.

The biomass and extra-cellular polymeric substances (EPSs) and sulfide precipitates were determined by filtering 5 mL samples through a dried and pre-weighed 0.22 μm membrane filter. The filter was then dried at 50°C for 5 days and re-weighed. The weight of total suspended solid (g/l) was then obtained by subtracting weight of fresh filter from weight of dried used filter.

RESULTS AND DISCUSSION

Batch Biokinetic Studies: In this study, sulfate was maintained at high concentration throughout the growth period, and since its value was much larger than that of K_{Sul}, which has been reported to be as low as 0.5-1.8 mg/l by Okabe *et. al.*(1992), the rate of the reaction was considered to be independent of sulfate concentration. Also, the decay term was neglected since high substrate concentrations were present for biomass growth. Hence, the rate of acetate reduction by SRBs can be written as follows:

$$r_{Ac} = -\frac{\mu_m C_{Ac}}{K_{Ac} + C_{Ac}} X \frac{1}{Y_{o/Ac}} \quad (1)$$

where, r_{Ac} is rate of substrate (acetate) utilization(g/l. hr); C_{Ac} and C_{sul} are concentration of acetate and sulfate (g/l), respectively; μ_m is maximum specific growth rate(hr^{-1}); $Y_{o/Ac}$ is the overall yield coefficient (g biomass/g acetate consumed) due to acetate utilization; X is the concentration of biomass (g/l); K_{Ac} and K_{Sul} are the Monod half-saturation coefficients for acetate and sulfate, respectively.

Biomass concentration can be written as follows:

$$X = X_0 + Y_{o/Ac} \times (C_{Ac0} - C_{Ac}) \quad (2)$$

where X_0 and C_{Ac0} are initial biomass and acetate concentrations, respectively.

Experimental results showed that SRB can not completely outcompete Methogen Producing Bacteria (MPB) unless sulfate concentration is approximately twice the acetate concentration (S/C ratio is larger than 2) in the batch reactor. As shown in Figures 1 and 2, under same pH and temperature, when S/C is smaller than 2, both MPB and SRB consumed acetate so that molar ratio of acetate consumed/sulfate reduced is greater than 1 while in the case of S/C larger than 2, MPB was completely inhibited by high concentration of sulfate and only SRB used acetate for sulfate reduction, such that the molar ratio of acetate consumed/sulfate reduced remained at 1. This phenomena is in accordance with the studies of Isa *et. al.*(1986) and Yoda *et. al.*(1987). They also found out that MPB consumed acetate much faster than SRB under limiting SO_4^{2-} conditions.

Kinetics of acetate-limited growth of sulfate reduction were determined at pH 5, 6, 7 and 8 and temperature 10°C, 25°C, 35°C, 45°C respectively. Estimates of μ_m and K_{Ac} were computed from equations (1) and (2). The standard error associated with each parameter was also calculated. The overall yield coefficient ($Y_{o/Ac}$)was defined here as the ratio of products formed to soluble acetate consumed. According to Okabe and Characklis (1991), in a biological reaction, products included cells, extracelluar polymeric compunds and sulfide precipitates. To determine the cell yield coefficient precisely, further identification of solid products is required. The highest maximum growth rate of $0.0121h^{-1}$ was obtained at pH 7 and 35°C. Below or above pH 7 and 35°C, maximum growth rate decreases. The lowest half-saturation coefficient K_{Ac}:1.4g/l was obtained at pH 7 and 25°C. At 10°C, K_{Ac} was much higher than those at higher temperature. The pattern of overall yields was the same as that of maximum growth rate. The lowest overall yield was 0.097g/g obtained at pH 5 and 10°C and the highest was 0.227g/g at pH 7 and 35°C.

The overall yield coefficients over acetate $Y_{o/Ac}$ were determined by experiments to be in the range of 0.097-0.227g total solid/g acetate in the pH range from 5 to 8 and in the temperature range 10°C to 45°C. Middeleton and Lawrence(1977) reported an overall yield coefficient 0.065g cells/g acetic acid consumed for mixed SRB culture utilizing acetic acid as limited carbon source in mixed continuous flow systems at the temperature 20, 25 and 31°C. Ingvorsen *et. al.*(1984) observed that an average overall growth yield of 0.0316 g(dry weight)/g acetate oxidized for Desulfobacter postgatei in the batch incubator at 30°C was obtained. Widdel and Pfennig(1981) reported an average growth yield of 0.032g cell dry weight per gram of acetate consumed for Desulfobacter postgatei in the batch reactor at 32°C. They also reported (1977) the growth yield of 0.036g-0.0486g cell dry weight/g acetate used for Desulfotomaculum acetoxidans in the batch system at 36°C.

Metal Precipitation Tests: Results calculated by the OLI computer program agree closely with experimental results. Copper (Cu^{2+}), Cadmium (Cd^{2+}), Cobalt (Co^{2+}), Lead (Pb^{2+}), Nickel (Ni^{2+}) and Zinc (Zn^{2+}) can be precipitated at low pH and ambient temperature, mainly due to low solubility of these metal sulfides. Manganese (Mn^{2+}) precipitates as sulfide at pH > 7.0 and Ferrous (Fe^{2+}) begins to precipitate at pH > 5.0. Other metals such as Aluminum (Al^{3+}) and Ferric (Fe^{3+}) cannot be precipitated as sulfides, since they form stable complexes. All the metals can be precipitated as hydroxides at 25°C when pH > 6.0, except Fe^{3+} and Al^{3+}, which begin to precipitate as hydroxides at pH > 3.0. When temperature is decreased to 10°C, metals such as Zn^{2+}, Fe^{3+} and Al^{3+} form stable complexes with OH^-, and hence cannot be precipitated as hydroxides. The solubility of some metal sulfides, such as Zn^{2+}, Pb^{2+} and Co^{2+} increases with decreasing temperature, and these metals cannot be precipitated as sulfides at 10°C.

Mine Reactor Tests: Figures 3 and 4 show the metal concentrations as a function of sample location (sample location 1 was located at the inlet of the acid mine water and sample location 6 was located at the top of the reactor near the exit point) and overall percent reduction for each metal. Results show that over 90% of Aluminum, Copper and Zinc were removed in the mine reactor at all temperatures. Removal of Iron (ferric) and Manganese increased with temperature. Manganese removal depended on the exit pH of the acid mine water. The average rates of sulfate reduction, obtained in the Mine reactor varied in the range of 270 mg/day at 10°C (overall sulfate reduction of 60%) to 356 mg/day at 25°C (overall sulfate reduction of 77.5%). The extent of sulfate reduction improved with temperature. The biokinetics of sulfate reduction determined from the batch reactor tests were used to simulate the performance of the Mine reactor, and good fits of the experimental data were obtained.

CONCLUSIONS

The biokinetic parameters determined in this study provide a quantitative basis for designing SRB reactors to reduce sulfate concentration in AMD treatment systems. Similar experimental methodology can be used to determine the biokinetics of sulfate reduction using other substrates, such as butyrate, lactate, or hydrogen gas. Results obtained from the Mine reactor demonstrate that injection of acetate solution can be used to *in-situ* bioremediate small remote mines, which would result in bulk precipitation of metal sulfides inside the mine shaft, resulting in a pH increase of the mine discharge. This provides a cost-effective solution for eliminating the deleterious effects of AMD discharges from remote mines. Metal precipitation studies showed that it is feasible to recover metals from large bodies of acid mine water, such as in Butte, Montana, by selectively precipitating the metals as either sulfides or hydroxides.

REFERENCES

Govind, R., Kumar, U.M., Puligadda, R., Antia, J. and Tabak, H. 1997. Biorecovery of Metals from Acid Mine Drainage, In *Emerging Technologies in Hazardous Waste Management 7*, Eds. D. W. Tedder, F.G. Pohland, Plenum Press, NY.

Ingvorsen, K. and JØrgensen, B. B., 1984, Kinetics of Sulfate Uptake by Freshwater and Marine Species of Desulfovibrio, ***Arch. Microbiol.***, Vol. 139, pp.61-66

Ingvorsen, K., Zehnder, A. J. B. and J. Ørgensen, B. B., 1984, Kinetics of Sulfate and Acetate Uptake by Desulfobacter postgatei, ***Appl. Environ. Microbiol.***, Vol. 47, pp.403-408

Isa, Z., Grusenmeyer, S. and Verstraete, W., 1986(a), Sulfate Reduction Relative to Methane Production in High-Rate Anaerobic Digestion: Technical Aspects, ***Appl. Environ. Microbiol.***, Vol. 51, pp.572-580

Isa, Z., Grusenmeyer, S. and Verstraete, W., 1986(b), Sulfate Reduction Relative to Methane Production in High-Rate Anaerobic Digestion: Microbiological Aspects, ***Appl. Environ. Microbiol.***, Vol. 51, pp.580-587

Middleton, A. C. and Lawrence, A. W., 1977, Kinetics of Microbial Sulfate Reduction, ***J. Water Pollut. Control Fed.*** Vol. 49, pp. 1659-1670

Okabe, S. and Characklis, W. G., 1992, Effects of Temperature and Phosphorous Concentration on Microbial Sulfate Reduction by Desulfovibrio Desulfuricans. ***Biotechnology and Bioengineering***, Vol. 39, pp. 1031-1042

Okabe, S., Nielsen, P. H., Characklis, W. G., 1992, Factors Effecting of Microbial Sulfate Reduction by Desulfovibrio Desulfuricans in Continuous Culture: Limiting Nutrients and Sulfide Concentration. ***Biotechnology and Bioengineering***, Vol. 40, pp. 725-734

Reis, M. A. M., Almeida, J. S., Lemos, P. C. and Carrondo, M. J. T., 1992, Effect of Hydrogen Sulfide on Growth of Sulfate Reducing Bacteria. ***Biotechnology and Bioengineering***, Vol. 40, pp. 593-600

Schauder, R., Eikmanns, B., Thauer, R. K., Widdel, F. and Fuchs, G., 1986, Acetate Oxidation to CO_2 in Anaerobic Bacteria Via a Novel Pathway Not Involving Reactions of the Citric Acid Cycle. ***Archives of Microbiology***, Vol. 145, pp. 162-172

Traore, A. S., Hatchikian, C. E., Gall, J. L. and Belaich, J. P., 1982, Microcalorimetric Studies of the Growth of Sulfate-Reducing Bacteria: Comparison of the Growth Parameters of Some Desulfovibrio Species, ***J. Bacteriol.***, Vol. 149, pp. 606-611

Traore, A. S., Hatchikian, C. E., Belaich, J. P. and Gall, J. L., 1982, Microcalorimetric Studies of the Growth of Sulfate-Reducing Bacteria: Energetics of Desulfovibrio Vulgaris Growth, ***J. Bacteriol.***, Vol. 145, pp. 191-199

Widdel, F. and Pfennig, N., 1977, A New Anaerobic, Sporing, Acetate-Oxidizing, Sulfate-Reducing Bacterium, Desulfotomaculum(emend.) Acetoxidans, ***Arch. Microbiol.***, Vol. 112, pp. 119-122

Widdel, F. and Pfennig, N., 1981, Studies on Dissimilatory Sulfate-Reducing Bacteria That Decompose Fatty Acids, ***Arch. Microbiol.***, Vol. 129, pp. 395-400

Widdel, F. and Pfennig, N., 1982, Studies on Dissimilatory Sulfate-Reducing Bacteria That Decompose Fatty Acids II Incomplete Oxidation of Propionate by Desulfobulbus Propoinicus Gen. Nov., Sp. Nov., ***Arch. Microbiol.***, Vol. 131, pp. 360-365

Yoda, M., Kitagawa, M. and Miyaji, Y., 1987, Long Term Competition between Sulfate Reducing and Methane Producing Bacteria for Acetate in Anaerobic Biofilm, ***Water Research***, Vol.21, pp. 1547-1552.

Figure 1. Concentration Profile of Acetate and Sulfate in the Batch Reactor at pH = 7.0 and 35°C, when the Sulfate / Acetate Concentration Ratio > 2.0.

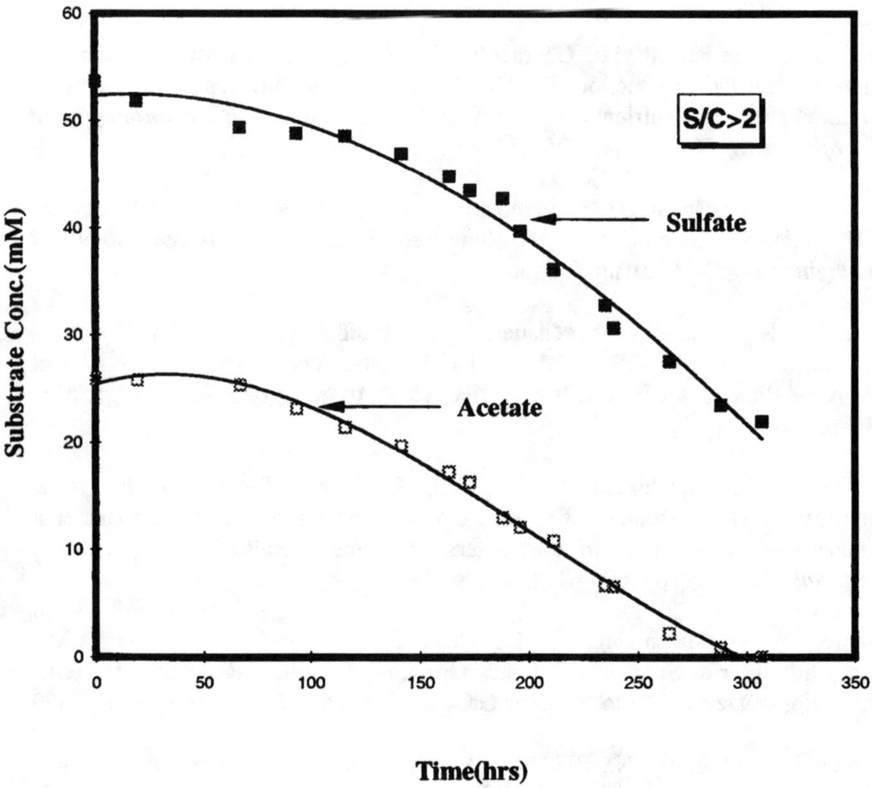

Figure 2. Concentration Profile of Acetate and Sulfate in the Batch Reactor at pH = 7.0 and 35°C, when the Sulfate / Acetate Concentration Ratio < 2.0.

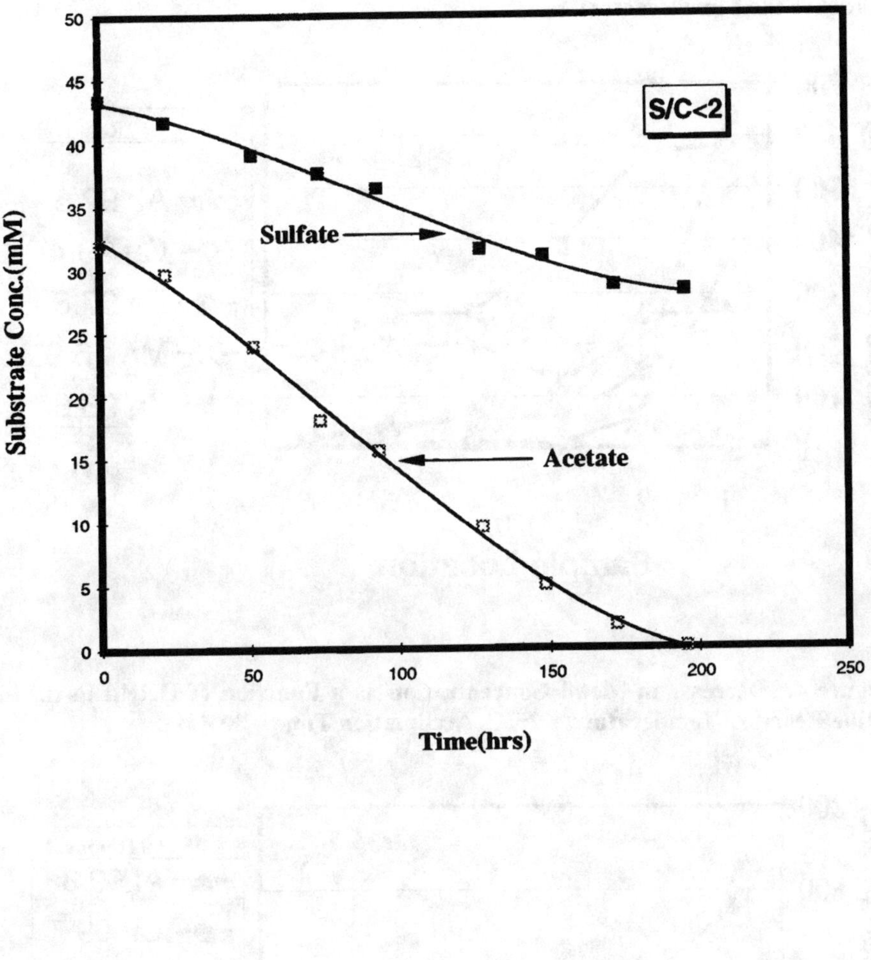

Figure 3. Decrease in Metal Concentrations as a function of Height in the Mine Reactor. Temperature = 10°C; Acclimation Time = 30 days. (Sample Location 1 is located at the Inlet and Sample Location 6 is at the Outlet of the Mine Reactor)

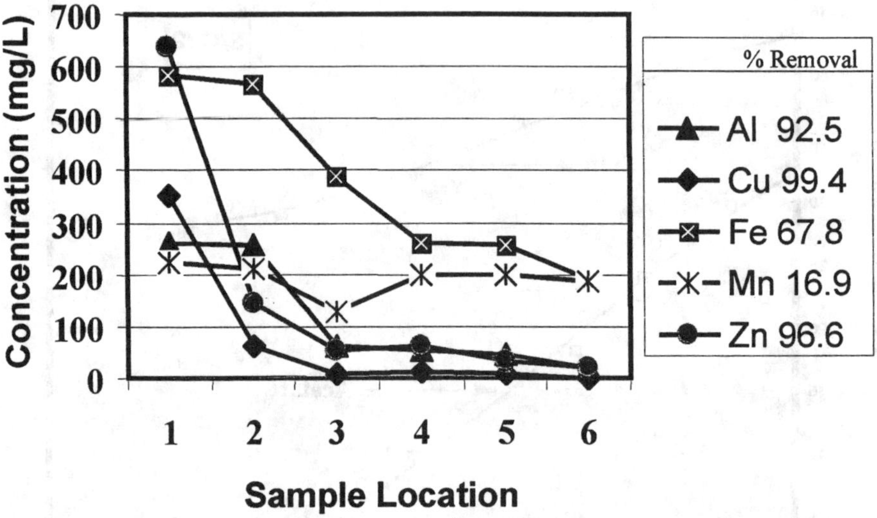

Figure 4. Decrease in Metal Concentration as a Function of Height in the Mine Reactor. Temperature = 25°C, Acclimation Time = 30 days.

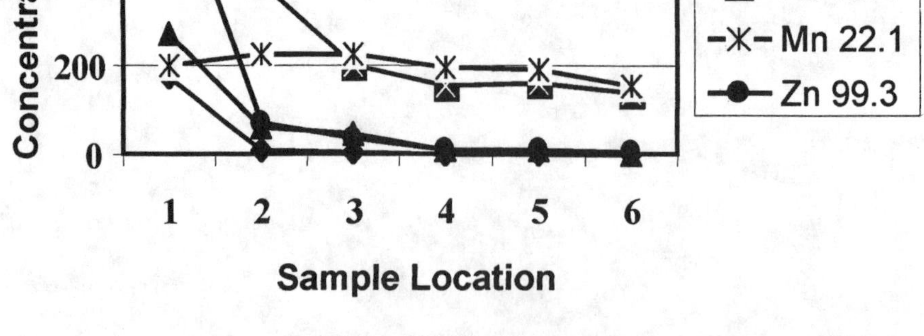

USE OF VEGETABLE OIL IN A PILOT SCALE DENITRIFYING BIOBARRIER

W. J. Hunter (USDA-ARS, Fort Collins, Colorado, USA)

ABSTRACT: Nitrate is a common groundwater contaminant. My studies suggest that biobarriers containing innocuous vegetable oils can be used to remove nitrate from groundwater. A sand tank was used as a scale model of an aquifer. Near the center of the tank was an area that contained 153 kg of sand coated with 1.5 kg of soybean oil that served as a denitrifying biobarrier. Over a 30 week period 33,400 liters of water containing 20 mg/L nitrate-N was pumped through the barrier. For the first 10 weeks of the study almost all of the nitrate was removed by the barrier, by week 15 the efficiency of the barrier was about 50%, and by week 30 almost no nitrate was removed. Denitrifying biobarriers containing innocuous vegetable oils could be used to protect water from nitrate contamination or treat contaminated water.

INTRODUCTION

Though high levels of nitrate contamination can occur from natural sources (Edmunds and Gaye, 1997) groundwater nitrate-N concentrations over ~3 mg/L usually indicate anthropogenic contamination (Foster et al., 1982; Kross et al., 1993). Agriculture is an important contributor to groundwater nitrate problems and in agricultural regions groundwaters often exceed the EPA drinking water standard of 10 mg/L nitrate-N (U.S. Public Health Service, 1962). The nitrate problem continues to grow as nitrates from fertilizers and animal wastes deposited years ago move slowly from the overlying soils into the water table (Gormly and Spalding, 1979; Spalding and Exner, 1993; Green and Shelef, 1994; Hamilton and Helsel, 1995) or move slowly within the aquifer (Hallberg, 1989).

Nitrate in drinking water is a health hazard to both humans and animals. Current recommendations are that water used for human consumption contain no more than 10 mg/L nitrate-N and water used for animals contain no more than 100 mg/L nitrate-N. Health problems that may be associated with nitrate in drinking water include methemoglobinemia, cancer and diabetes. Nitrate may accumulate in soils and waters when denitrification cannot occur. This is usually due to the absence of an appropriate electron donor that limits the activity of denitrifying microorganisms. Microorganisms capable of removing nitrate from water by denitrification are normally present in groundwater. Laboratory studies have shown that vegetable oil can serve as an electron donor, allowing denitrifying organisms to utilize the nitrates present in the groundwater as a term electron acceptor via denitrification (Hunter et al., 1997; Hunter and Follett, 1997). The overall result of denitrification is that vegetable oil and nitrate are converted to biomass, nitrogen gas, water, carbon dioxide, and, most importantly, that the nitrate content of the water is reduced. Vegetable oils differ from most other carbon substrates that have been used as electron donors for denitrification in that they are not miscible with water. Thus vegetable oils can be used to form a stationary zone of denitrification through which water can flow.

METHODS

A 100 cm long biobarrier containing soybean oil was placed in a 110 cm high x 200 cm long x 8.5 cm wide open top sand tank. The biobarrier was made by coating 153 kg of sand with 1.53 kg of soybean oil and filling a 100 cm section of the sand tank with the oil coated sand. The biobarrier served as a denitrifying zone removing nitrate biologically while allowing water to pass. A 20-cm-long zone of uncoated sand was placed in front of the biobarrier and a 70 cm long zone of uncoated sand placed behind the biobarrier. A moderately hard reconstituted water (Greenberg et al., 1992) supplemented with phosphate and 20 mg/L nitrate-N was pumped through the sand tank at an average flow rate of 144 L/day. Pumped water passed through the biobarrier while flowing horizontally between a screened influent pipe placed at one end of the sand tank and a screened effluent pipe placed at the other end of the tank. The temperature of the tank was maintained at 15°C. No attempt was made to deoxygenate the supply water.

RESULTS AND DISCUSSION

Over a 30 week period 33,400 L of water containing 20 ppm nitrate-N was pumped through the sand tank and across the denitrifying biobarrier. The initial efficiency of nitrate removal was ~100% but this efficiency declined with time such that by week 15 about 50% of the nitrate in the influent was being removed and by week 30 almost none of the influent nitrate was removed. Overall, during the 30 week study, 39% or 261 g of nitrate-N was removed from the effluent water. Thus each gram of oil removed 171 mg of nitrate-N from the oxygenated influent water. This efficiency is 76% of that which was observed previously with small laboratory soil columns where one gram of oil removed 224 mg of nitrate-N (Hunter et al., 1997). The loss of organic material (discussed below) would explain some of the loss of efficiency.

Nitrite Accumulation. Nitrite levels in the effluent water remained below 0.4 mg/L nitrite-N throughout the study. In previous studies with small soil columns nitrite accumulation has sometimes been observed as the

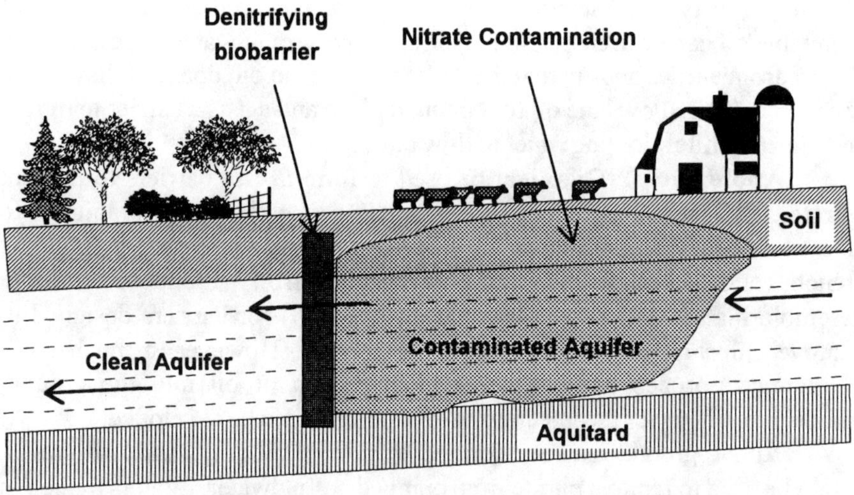

FIGURE 1. Hypothetical use of a vegetable oil based biobarrier to protect an uncontaminated portion of an aquifer from a portion contaminated by nitrate formed from animal wastes.

vegetable oil was consumed by microbial activity and carbon source (oil) availability begin to limit microbial activity (Hunter et al., 1997) but this was not observed with the sand tank.

Hydraulic Conductivity. Biomass accumulations often interfere with the flow of water in denitrifying reactors. Hydraulic conductivity between two piezometer ports on each side of the denitrifying zone was monitored and found to remain relatively constant (between 0.1 and 0.2 cm sec^{-1}) through the study.

COD: The chemical oxygen demand (COD) of the effluent water was high, averaging 78 and ranging between 50 and 130 mg oxygen/L, for the first 10 weeks of the study. These results indicate a loss of organic material, either as oil or biomass, from the tank during this period. It is likely that a larger sand zone behind the denitrifying barrier would have trapped some of this organic material. This trapped organic material could serve as an electron donor for denitrification, thus improving the efficiency of the system. Loss of organic material would be less of a problem under *in situ* conditions. As would be expected with waters this high in COD, and as is usually observed with biological denitrification reactors, the effluent waters were turbid and high in total suspended solids (TSS). COD was much lower during the last 20 weeks of the 33 week study, averaging 5.0 and ranging between 0 and 12 mg oxygen/L.

Conclusions. These studies demonstrate the principles whereby a simple denitrifying zone composed of sand, gravel, and small amounts of vegetable oil can be used to remove nitrate from groundwater. The approach differs from earlier ones in that the carbon source, an oil, does not dissolve in the water. This allows the oil to remain in place and act as a barrier to nitrate movement while allowing water to flow through.

Applications. Oil might be used to form *in situ* barriers to prevent nitrate from spreading from a contaminated section of an aquifer to uncontaminated areas (Figure 1). Barriers might be formed by digging trenches and backfilling them with sand coated with oil. This approach would resemble that described by Gavaskar et al. (1998) for barriers designed to remove chlorinated solvents from groundwater. However, a much more economic approach might involve the injection of oil into the ground, although methods for the injection of oil would need to be developed.

Above ground reactors, formed by mixing oil with sand and gravel, might be used to remove nitrate from pumped groundwater. Waters treated in above ground reactors would require secondary treatment if intended for consumption (Dahab and Sirigina, 1994; Green and Shelef, 1994; Hunter and Follett, 1997). Under *in situ* conditions high COD, turbidity, and TSS would be less of an issue as the aquifer itself would provide many of the functions of a secondary reactor and act to cleanse the denitrified water (Green and Shelef, 1994).

REFERENCES

Dahab, M. F., and S. Sirigina. 1994. "Nitrate Removal from Water Supplies Using Biodenitrification and GAC-sand Filter Systems." *Wat. Sci. Tech. 30*(9):133-139.

Edmunds, W. M., and C. B. Gaye. 1997. "Naturally High Nitrate Concentrations in Groundwaters from the Sahel." *J. Env. Qual. 26*(5):1231-1239.

Foster, S. S. D., A. C. Cripps, and A. Smith Carington. 1982. "Nitrate Leaching to Groundwater". *Phil. Trans. Rol. Soc. London. 296:*477-489.

Gavaskar, A. R., N. Gupta, B. M. Sass, R. J. Janosy, and D. O'Sullivan. 1998. *Permeable Barriers for Groundwater Remediation: Design, Construction, and Monitoring.* Battelle Press, Columbus, OH

Gormly, J. R., and R. F. Spalding. 1979. "Sources and Concentrations of

Nitrate-Nitrogen in Ground Water of the Central Platte Region, Nebraska." *Ground Water 17*(3):291-301.

Green M., and G. Shelef. 1994. "Treatment of Nitrate Contaminated Groundwater for Drinking Purposes." In U. Zoller (Ed.), *Groundwater Contamination and Control*, pp. 587-606. Marcel Dekker Publishers, New York, NY.

Greenberg, A. E., L. S. Clesceri, and A. D. Eaton (Eds.). 1992. *Standard Methods for the Examination of Water and Wastewater.* 18th ed. Washington, D.C.: Am. Pub. Health Assoc., Am. Water Works Assoc. and Water Environ. Fed.

Hallberg, G. R. 1989. "Nitrate in Ground Water in the United States." In R. F. Follett (ed), *Nitrogen Management and Ground-Water Protection,* pp. 35-74. Elsevier Science Publishing Co., New York, NY.

Hamilton, P. A., and D. R. Helsel. 1995. "Effects of Agriculture on Ground-Water Quality in Five Regions of the United States." *Ground Water 33*(2):217-226.

Hunter, W. J. and Follett, R. F. 1997. "Removing Nitrate From Groundwater Using Innocuous Oils: Water Quality Studies." In B. C. Alleman and A. Leeson (eds), *In Situ and On-Site Bioremediation: Volume 3,* pp. 415 - 420. Battelle Press, Columbus, OH.

Hunter, W. J., R. F. Follett, and J. W. Cary. 1997. "Use of Vegetable Oil to Stimulate Denitrification and Remove Nitrate from Flowing Water." *Trans. ASAE 40*(2):345-353.

Kross, B. C., G. R. Hallberg, D. R. Bruner, K. Cherryholmes, and J. K. Johnson. 1993. "The Nitrate Contamination of Private Well Water in Iowa." *Am. J. Public Health. 83*(2):270-272.

Spalding, R. F., and M. E. Exner. 1993. "Occurrence of Nitrate in Groundwater--A Review." *J. Environ. Qual. 22*(3):392-402.

U. S. Public Health Service. 1962. *Public Health Drinking Water Standards.* U. S. Public Health Publication # 956. U.S. Government Printing Office, Washington, D. C.

BIOREMEDIATION OF NITRATE CONTAMINATED SOILS USING INNOCUOUS OIL

W. J. Hunter (USDA-ARS, Fort Collins, Colorado, USA)

ABSTRACT: Agricultural operations that involve manure can cause the contamination of groundwater with nitrate. Innocuous vegetable oils might be used as an electron donor in a denitrifying biobarrier to prevent the downward movement of nitrate in contaminated soils. This study, at a site in northeastern Colorado, evaluated the movement of nitrate in soils beneath a small cattle pen and the use of innocuous vegetables oil to stimulate denitrification in those soils. Soil cores were collected to a depth of 160 cm beneath the pen. Soils contained up to 330 mg nitrate-N/kg with the highest levels occurring deeper (below 40 cm). However, denitrification activity was highest near the surface above these high nitrate concentrations. Adding oil as an electron donor stimulated denitrification in the deeper soils. In these pens denitrification activity near the surface has not stopped nitrate from moving into the deeper soil layers where denitrification activity was limited by the amount of electron donor (carbon) present. Over time nitrate has moved through these soils and into the aquifer below the pens. My results suggest that innocuous vegetable oils could be used in biobarriers to block the movement of nitrate into deeper soils. *In situ* applications could involve the injection of oil into the ground to prevent the seepage of nitrate from a feedlot, manure storage area or industrial site.

INTRODUCTION

Nitrate contamination of subsurface soils is a common problem in many parts of the United States. Once nitrates have moved below the root zone it is likely that their downward migration will continue and eventually result in contamination of the groundwater. The rate of nitrate movement from contaminated soils in the lower vadose zone into the water table is influenced by soil properties and climatic conditions. In dry climates with fine textured soils many decades might be required for nitrate to move into the water table, whereas under irrigation or moist climates nitrates present in thin coarse-textured soils can move rapidly into the water table (Pionke and Lowrance, 1991).

Nitrate contamination of groundwater is a concern because of the effects that nitrite may have on human health (National Academy of Sciences, 1978; Train, 1979; Kross et al., 1993). The USEPA (1973) has established a drinking water standard of 10 mg/L for human consumption and the National Academy of Sciences-National Research Council (Bruning-Fann and Kaneene, 1993) has recommended that drinking water for farm animals contain no more than 100 mg/L nitrate-N. In many parts of the United States groundwaters exceed these recommendations. In Iowa a statewide survey of rural wells conducted in 1988 and 1989 shows that 18% of the private drinking-water wells exceeded the EPA standard of 10 mg/L nitrate-N and in Kansas a statewide survey showed that 28% of tested wells exceeded the EPA standard (Kross et al., 1993). A study by Hamilton and Helsel (1995) of five regions in the US found that in central and western Connecticut, 12%; in the high plains of Nebraska, 46%; in south-central Kansas, 17%; in Long Island NY, 27%; and in the Delmarva Peninsula of Delaware and Maryland, 33% of well water samples exceeded the EPA water standard. In northeastern Colorado a survey found that 70% of wells sampled exceeded the EPA standard (Schuff, 1992; Wylie et al., 1994; Wylie et al., 1995). These are but a few of the many studies that show that nitrate contamination of groundwater is a major problem. Due to the costs associated with removing nitrate from drinking water communities have been forced to abandon contaminated groundwater sources and seek cleaner sources of water (Schuff, 1992; Spalding and Exner, 1993).

Microbial denitrification can remove nitrate from deep soils and groundwaters. While a number of factors affect this process it is most often the availability of an electron donor that limits the rate at which this process proceeds (Myrold and Tiedje, 1985 and others). The addition of a carbon source to contaminated soils or aquifers may be the most economic approach for removing nitrate (Streile et al., 1991; Green and Shelef, 1994).

While nitrates may come from many sources, in Colorado nitrates derived from animal wastes are a common problem. Pervious studies have shown that innocuous oils can be used to form permeable biobarriers for removing nitrate from flowing groundwater (Hunter et al., 1997). The present study investigated the distribution of nitrates in subsurface soils beneath a cattle holding pen and the use of innocuous oils to stimulate denitrification in these soils.

METHODS

Soil cores were collected to a depth of 160 cm beneath a cattle pen located on the USDA-ARS Central Plains Experimental Range in northeastern Colorado and transported to the lab for analysis and study. Denitrifying enzyme activity (DEA) present in the soil cores was determined by the method of Tiedje (1982). Serum bottles (123 ml) containing soil, buffer and a helium atmosphere (spiked with 10 ml acetylene) and closed with a rubber serum stopper were used as the DEA reactors. DEA activity was determined by the amount of N_2O that accumulated in the atmosphere of the bottle. N_2O was analyzed with a gas chromatograph equipped with thermal conductivity and electron capture detectors as described in Hunter et al. (1997).

Nitrate and chloride were extracted from the soil samples with water [1:5 (soil:water)]. Soil samples were shaken on a wrist action shaker for 30 min, centrifuged for 15 min at 16,000 g and the supernatant fluid saved. Nitrate and chloride in the soil extracts were determined by high pressure liquid

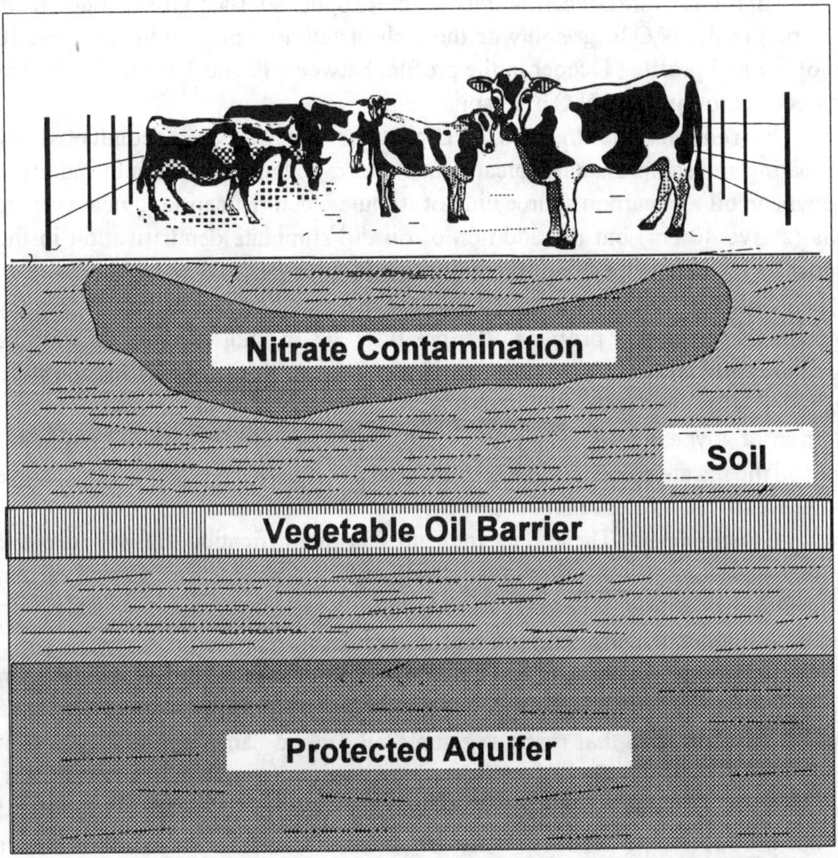

FIGURE 1. Possible use of a vegetable oil barrier to prevent nitrate from migrating into an aquifer.

chromatography. The UV detection method used for nitrate is described in Hunter et al. (1997). Chloride was determined using conductivity detection, a Hamilton PRP-X100 anion exchange column, and 4mM *p*-hydroxybenzoic acid (pH 8.5) as the elution buffer as described in Hamilton application method #73 (Hamilton Company, 1993).

Bioreactor incubations were conducted as described in Hunter et al. (1997).

RESULTS AND DISCUSSION

Soil Analysis. The soils were structureless and contained between 1 and 330 mg/L nitrate-N. Highest nitrate (and chloride) levels occurred in soils collected from between 40 and 160 cm depth where nitrate-N levels ranged from 66 to 330 mg/L. DEA was highest in the upper 40 cm of the soil profile and above the area of high nitrate concentrations. In these upper soils activities ranged from 35 to 62 μmoles N_2O h^{-1} g^{-1} soil with the highest activities present in the upper 10 cm of the soil profile. Deeper in the profile, between 40 and 160 cm, DEA was between 0.1 to 8 μmoles N_2O h^{-1} g^{-1} soil.

Bioreactor Incubations. Bioreactor studies were conducted by incubating soil samples anaerobically for 1 week. Results show that the addition of soybean oil as a carbon source did not stimulate denitrification in near surface soils (above 40 cm) but the addition of oil did stimulate denitrification in the deeper soils where large amounts of nitrates were present.

Column Incubations. Soil column studies were conducted by reconstructing the soil profile in 2.6 cm ID x 1.8 m long columns. Columns containing soil with moisture contents of 40, 50, 60 or 70% pore filled space were evaluated. Columns were incubated at 15°C for 30 days. Results showed that the addition of soybean oil (1.13g oil kg^{-1} soil) to deeper soils (40 to 120 cm deep) reduced the nitrate content of these soils by an average of 47 ± 7%. Soil moisture content had little influence on the amount of nitrate removed.

Conclusions. These data indicate that denitrification activity near the surface was limited by nitrate availability and not by electron donor (carbon substrate) availability. However in deeper soils, below 40 cm, it was electron donor availability that limited denitrification activity.

Laboratory bioreactor and column studies show that innocuous oil can function as an electron donor for denitrification in these deeper soils. There are a number of properties that make vegetable oil a good candidate as an electron donor source for denitrification. Oils are inexpensive, have a high energy content, are readily degradable by a large population of microorganisms, are not toxic and do not produce toxic products when degraded, and are not miscible in water (Hunter et al., 1997; Hunter and Follett, 1997). *In situ* denitrification barriers based on innocuous oils, such as vegetable oil, should provide a simple approach for protecting groundwater from nitrate contamination. Since innocuous oils are liquids it should be possible to inject the oil into the ground to form biobarriers (Figure 1) to prevent the movement of nitrate into an aquifer or to remediate waters that are contaminated with nitrate.

REFERENCES

Bruning-Fan, C. S. and J. B. Kaneene. 1993. "The Effects of Nitrate, Nitrite, and N-nitroso Compounds on Animal Health." *Vet. Hum. Toxicol. 35*:237-253.

Green M., and G. Shelef. 1994. "Treatment of Nitrate Contaminated Groundwater for Drinking Purposes." In U. Zoller (Ed.), *Groundwater Contamination and Control*, pp. 587-606. Marcel Dekker Publishers, New York, NY.

Hamilton Company, 1993. "Application #73: Seven Anions" In *The Hamilton*

HPLC Application Handbook, p. 70. The Hamilton Company, Reno, NV.

Hamilton, P. A., and D. R. Helsel. 1995. "Effects of Agriculture on Ground-Water Quality in Five Regions of the United States." *Ground Water 33*(2):217-226.

Hunter, W. J. and Follett, R. F. 1997. "Removing Nitrate From Groundwater Using Innocuous Oils: Water Quality Studies." In B. C. Alleman and A. Leeson (eds) *In Situ and On-Site Bioremediation: Volume 3*, pp. 415-420. Battelle Press, Columbus, OH.

Hunter, W. J., R. F. Follett, and J. W. Cary. 1997. "Use of Vegetable Oil to Stimulate Denitrification and Remove Nitrate from Flowing Water." *Trans. ASAE 40*(2):345-353.

Kross, B. C., G. R. Hallberg, D. R. Bruner, K. Cherryholmes, and J. K. Johnson. 1993. "The Nitrate Contamination of Private Well Water in Iowa." *Am. J. Public Health. 83*(2):270-272.

Myrold, D. D., and J. M. Tiedje. 1985. "Establishment of Denitrification Capacity in Soil: Effects of Carbon, Nitrate and Moisture." *Soil Biol. Biochem. 17*(6):819-822.

National Academy of Sciences. 1978. Nitrates: An Environmental Assessment. Washington, D. C.

Pionke, H. B., and R. R. Lowrance. 1991. "Fate of Nitrate in Subsurface Drainage Waters." In R. F. Follett, D. R. Keeney, and R. M. Cruse (eds), *Managing Nitrogen for groundwater Quality and Farm Profitability*, pp. 237-257. Soil Science Society of America, Madison, WI.

Spalding, R. F., and M. E. Exner. 1993. "Occurrence of Nitrate in Groundwater--a Review. *J. Environ. Qual. 22*(3):392-402.

Tiedje, J. M. 1982. "Denitrification." In L. A. Page, R. H. Miller and D. R. Keeney (eds) *Methods of Soil Analysis, Part 2. Chemical and Microbiological Properties*. pp. 1011- 1026. Agronomy Monograph #9 (2nd edition). Madison, WI.

Train, R. E. 1979. *Quality Criteria for Water.* Castle House Publications, London.

Schuff, S. 1992. "Nitrates Can Leach, But They Can't Hide." *Colorado Rancher and Farmer. 46*(14):6-12.

Streile, G. P., J. W. Cary, and J. K. Fredrickson. 1991. "Innocuous Oil and Small Thermal Gradients for In Situ Remediation." In *Proceedings of the National Research and Development Conference on the Control of Hazardous Materials*, pp. 32-37. Anaheim, CA.

U.S. Environmental Protection Agency. 1973. *Nitrogenous Compounds in the Environment.* U.S. Government Printing Office, Washington, D.C. EPA-SAB-73-001.

Wylie, B.K., Shaffer, M.J., Brodahl, M.K., Dubois, D. and Wagner, D.G. 1994. "Predicting Spatial Distributions of Nitrate Leaching in Northeastern Colorado". *J. Soil and Water Cons. 49*(3):288-293.

Wylie, B.K., Shaffer, M.J. and Hall, M.D. 1995. "Regional Assessment of NLEAP NO_3-N Leaching Indices." *Water Resources Bulletin. 31*(3):399-408.

IN SITU DENITRIFICATION : A FIELD DEMONSTRATION

H. Eric Nuttall, Lijun Deng, Abdesselam Abdelouas, Werner Lutze
The University of New Mexico, Albuquerque, New Mexico, U.S.A.

ABSTRACT: An in situ ground water denitrification process was developed using a series of increasingly larger scale tests, starting from microcosm scale experiments, through pilot scale tests, and finally a field demonstration. Indigenous soil and ground water denitrifying bacteria were stimulated using sodium acetate as carbon substrate and trimetaphosphate as nutrient. In situ denitrification was successfully demonstrated using a push-pull test at a New Mexico site located in Albuquerque' s South Valley. Ground water parameters of pH, Eh, conductivity, concentrations of dissolved oxygen, nitrate, nitrite, acetate, sulfate, and bromide were measured. Results showed that the high toxic levels of nitrate (100 mg/L nitrate nitrogen) were metabolically transformed into harmless nitrogen gas within three-five days, and the nitrate concentration was reduced to below 1 mg/L. The final pH, Eh, and conductivity values of ground water were not significantly changed during the process.

INTRODUCTION

Nitrate contamination in drinking water poses serious health effects to both humans and animals. Infants are most seriously impacted. Blue baby syndrome (methemoglobinemia) is potentially fatal when infants (from birth to six months) drink contaminated water at nitrate levels above MCL (10 mg/L nitrate-nitrogen). Other report diseases caused by drinking water polluted with nitrate include: cancer, disruption of thyroid function, and birth defects. Nitrate contamination in ground water is most commonly caused by over fertilization of agricultural crops, improper disposal of animal and human waste, and industrial liquid waste.

Remediation of nitrate contaminated ground water and soils can occur under certain conditions by natural bioremediation. Nature has evolved indigenous denitrifying bacteria capable of converting nitrates found in subsurface environments, i.e., soils and ground water, into harmless nitrogen gas (Payne, 1982); however, the process is often very slow requiring decades or centuries to complete due to the lack of a carbon energy source for bacterial growth. Enhanced or accelerated bio-denitrification is accomplished by stimulating indigenous denitrifying bacteria through the addition of a suitable carbon energy source (McCarty et al., 1969). Carbon substrates such as methanol, ethanol, acetate, and sugar can significantly enhance denitrification rates by serving as electron donor and energy supply for the indigenous bacteria, while nitrate is the electron acceptor. In the metabolic denitrification process nitrate is transformed into final product nitrogen gas via a multistep chemical reduction: $NO_3^- \rightarrow NO_2^- \rightarrow NO\ (g) \rightarrow N_2O\ (g) \rightarrow N_2\ (g)$ (Knowles, 1982). Usually it is found that nitrite is the most significant intermediate while NO and N_2O are short-lived. In practice the dissolved oxygen is also an electron acceptor and must be removed by bacteria prior to denitrification. The stoichiometric chemical reaction for deoxygenation is described in equation (1), in which we assume that 30% of carbon is consumed for the cell synthesis (Deng, 1998). The stoichiometric relation describing denitrification using acetate was experimentally derived (equation 2) (Lu, 1998), and indicates that most of the carbon source (90%) is used to provide energy for the bacteria rather than biomass production.

$$\mathbf{O_2 + 0.816\ CH_3COO^- + 0.103\ NO_3^- + 0.964\ H^+ \rightarrow 0.103C_5H_7O_2N + 1.207\ CO_2 + 1.413\ H_2O} \quad (1)$$

$$\mathbf{NO_3^- + 0.712\ CH_3COO^- \rightarrow 0.485\ N_2 + 0.03\ C_5H_7O_2N + 1.273\ HCO_3^- + 0.106\ H_2O + 0.44\ OH^-} \quad (2)$$

where $C_5H_7O_2N$ represents the cell formula.

Because enhanced denitrification is rapid and can be applied in situ, it appears likely to be a cost-effective technology for treating ground water and soils. Many denitrification studies (Kappelhof, et al., 1992, Green et al., 1994) have investigated ex situ bio-denitrifying reactor systems; however, only a few studies have investigated in situ bio-denitrification (Mercado, et al., 1988; Hamon and Fustec, 1991).

Objective. The objective of this study is to determine the feasibility of in situ biological denitrification at a site in New Mexico (Nuttall, et al., 1997, Nuttall, 1997, Deng et al., 1998, Abdelouas et al., 1998). The approach was to scale up the process in a series of batch microcosm experiments in closed reactors ranging in volume from 0.16 L to 4542 L (1200 gal). Following successful completion of the ex situ tests, an in situ denitrification experiment was performed.

Site Description. The in situ denitrification site located in Albuquerque, New Mexico's South Valley is a 40 year old nitrate plume covering an area of about 500 acres and a volume of 6.4 billion liters. Ground water contamination was caused by over fertilization at a vegetable farm in the 1950s. The water table is at a depth of 22 meters and the top of the aquifer contains a 10-meter nitrate contamination zone. The plume is moving very slowly in a sandy/loam soil aquifer with a hydraulic conductivity of about 10^{-3} cm/sec. The average nitrate nitrogen concentration in the ground water is 90 - 110 mg/L. The ground water pH is 7.4-7.7 with a temperature of 20°C. At this site in 1981, a blue baby syndrome incident was reported.

EXPERIMENTS AND RESULTS

Ex Situ Denitrification Scale-up Tests. The biotreatability of ground water was first investigated in batch experiments in 160 mL closed serum bottles. The ground water was collected from a single well located at the Tri-Tech site in Albuquerque's South Valley. The required amount of sodium acetate was determined using equations (1) and (2). In practice, sodium acetate was added to provide a C/N ratio of 1.42 – 2.00. Sodium trimetaphosphate(TMP) was added to give a concentration of 15-20 mg/L. Further scale-up tests were performed in a

208 liter (55 gallon) drum and a 4542 liter (1200 gallon) tank using the same carbon substrate and phosphorus nutrient. Reaction times for complete nitrate removal ranged from 7-15 days as shown in Table 1.

Table 1. Ex situ denitrification scale-up tests using indigenous ground water bacteria (acetate used as carbon substrate, TMP as phosphorus nutrient)

Reactor Type	Volume (L)	Scale Ratio	Total Nitrogen (mg/L) Initial // Final	Time (days) 20°C	Denitrification Efficiency(%)
Serum Bottle	0.12	1	99.4 // <0.1	7	100
Drum	200	1,667	101.6 // 1.1	7	99
Tank	4200	35,000	115.5 // <0.1	15	100

In Situ Denitrification Field Test. The schematic diagram for the push-pull in situ field test is illustrated in Figure 1. The test was designed to keep the reacting ground water between the two monitoring wells. About 4200 L of contaminated ground water was pumped out from an extraction well and mixed with sodium acetate and TMP in a mixing tank. A bromide tracer was used to follow dilution of the reacting ground water. The amended ground water with bromide as tracer was recharged into the injection well at a flow rate of 6 L/min. The ground water from injection and monitoring wells was sampled on a daily basis during denitrification. The post-sampling was carried out every 15 days over a period of two months. The groundwater parameters of pH, conductivity, TDS, DO, and Eh were measured in the field by a YSI 6920 water quality monitor installed in the flow cell as shown in Figure 1. Water samples were measured in the lab using a Dionex DX 500 chromatography system for anions: acetate, nitrate, nitrite, and sulfate. The concentration of bromide was measured using an ATI Orion Model 9635 *ionPlus* Series Bromide Electrode.

Figure 2 shows the concentrations of acetate, nitrate, and nitrite versus time for samples taken from the injection well during denitrification. All traces of both nitrate and nitrite intermediate were removed in five days. The nitrate was metabolically transformed to nitrogen gas, and the final concentration was less than 1 mg/L. Table 2 shows a comparison of average denitrification rates for three different process scales indicating that in situ average denitrification is nearly three times the average rate of ex situ tests. The faster in situ denitrification is likely due to the significantly greater bacterial concentration in the aquifer matrix as compared to the bacteria concentrations in water samples.

Figure 3 shows the concentrations of acetate, nitrate, nitrite and bromide versus time for water samples from the injection well over the two month monitoring period. After the initial denitrification reaction was completed, the excess acetate was eventually consumed by further denitrification resulting from mixing of the surrounding contaminated ground water. This was confirmed by the decrease of the tracer bromide. During the second reaction period, the nitrate concentration remained low until all the excess acetate was consumed.

The long-term monitoring of groundwater parameters of Eh, pH, conductivity, and dissolved oxygen were documented in Deng, 1998. Results also indicated that the Eh, pH, and conductivity values returned to background levels following the push-pull field demonstration.

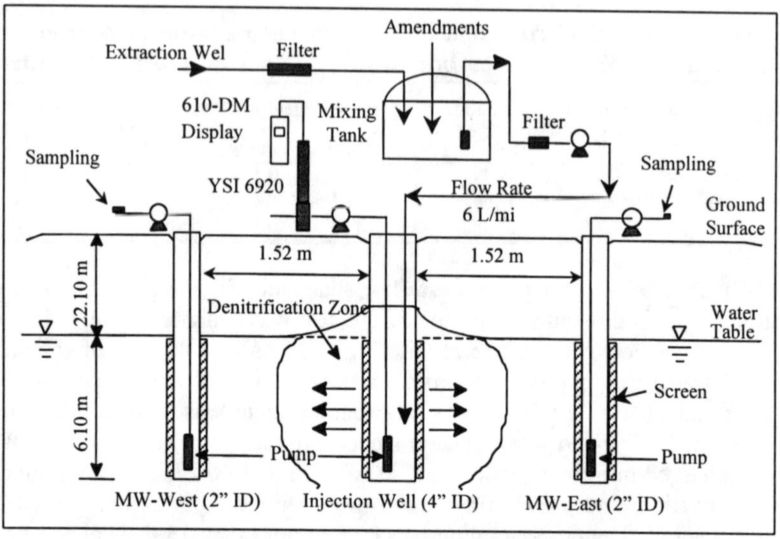

Figure 1. Design of in situ denitrification test at Tri-Tech site, Albuquerque' s South Valley (not drawn to scale).

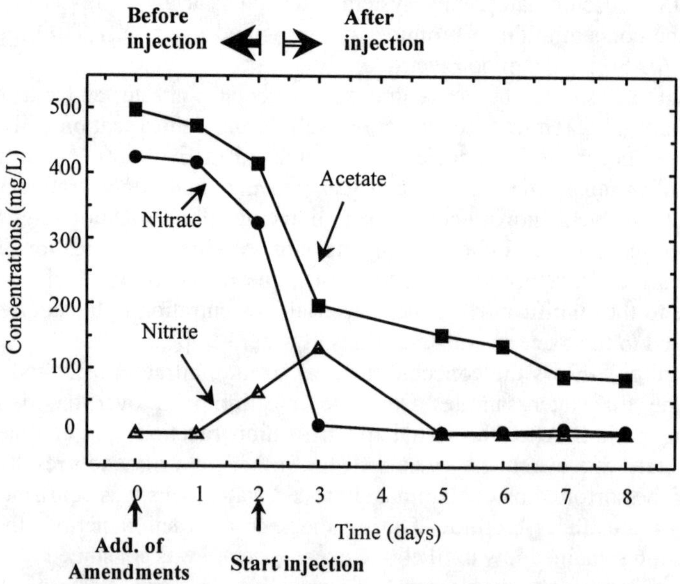

Figure 2. Reaction concentration curves resulting from a short term monitoring during in situ denitrification test.

Table 2. Comparison of average denitrification rates using acetate and TMP as amendments.

Experiments	Temperature (°C)	Denitrification Time (days)	Average Denitrification Rate (mg N /L/day)
serum bottle test w/o sand	25	18	5.4
1200 gal tank test	20	15	6.9
In situ field test	20	5	19.0

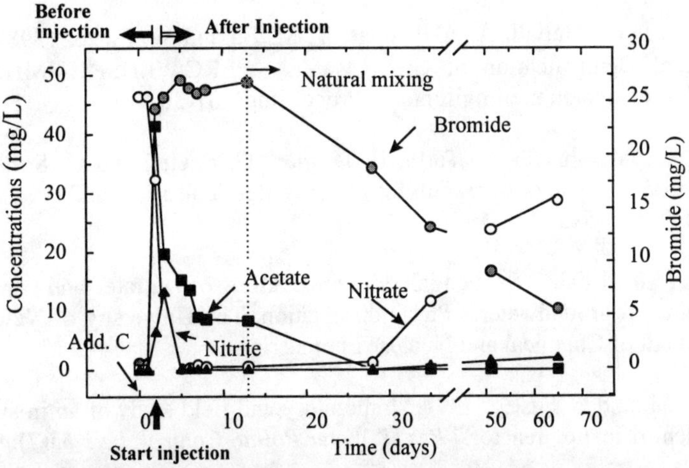

Figure 3. Reaction concentration curves measured during in situ biodenitrification test.

CONCLUSIONS

In situ field denitrification performed very well resulting in complete nitrate removal and conversion to harmless nitrogen gas in less than five days. The approach of scaling-up the process in steps was very useful to confirm reactions rates and to show that reactor size is not important as long as there is good mixing of the amendments with the ground water. Because of the positive results, a larger scale continuous in situ demonstration is planned. The upcoming continuous in situ tests will evaluate costs, operation parameters/design, and address the issue of potential biofouling. Though the overall costs for in situ bio-denitrification of ground water is still being assessed, our evaluation indicates that the amendment (sodium acetate and trimetaphosphate) costs will be about $0.64 per 1000 gal of ground water containing 20 mg/L nitrate nitrogen and 4 mg/L DO.

ACKNOWLEDGMENTS

The research on which this paper is based was financed in part by the U.S. Department of Energy through the New Mexico Waste-management Education and Research Consortium (WERC). The experiments and measurements were done at the Center for Radioactive Waste Management (CeRaM) at the University of New Mexico.

REFERENCES

Abdelouas, A., Y. M. Lu, W. Lutze, and H.E. Nuttall. 1998. "Reduction of U(VI) to U(IV) by indigenous bacteria in contaminated groundwater." *J. Contam. Hydrology. 35*: 217-233.

Deng, L. J. 1998. "In situ biological denitrification of groundwater." Master thesis. The University of New Mexico, Department of Chemical and Nuclear Engineering.

Deng, L., H. E. Nuttall, A. Abdelouas, Y. M. Lu and W. Lutze. 1998. "In Situ Biological Denitrification of Groundwater." WERC WRHSRC NMHWMS '98 Joint Conference on the Environment Proceedings. 31-36.

Green, M., M. Shnizer, S. Tarre, B. Bogden, G. Shelef, and C. Sorden. 1994. "Fluidizied bed reactor operation for groundwater denitrification." *Wat. Sci. Tech. 29*(10/11): 509-515.

Lu, Y. M. 1998. "Sequential bioremediation of nitrate and uranium in contaminated groundwater." Ph. D dissertation. The University of New Mexico, Department of Chemical and Nuclear Engineering.

Hamon, M. and E. Fustec. 1991. "Laboratory and field study of an in situ ground water denitrification reactor." *Res. J. Water Pollut. Control. Fed. 63*(7): 942-949.

Kappelhof, J. W. N. M., J. P. van Der Hoek, and W. A. M. Hijnen. 1992. "Experience with fixed-bed denitrification using ethanol as substrate for nitrate removal from groundwater." *Water Supply. 10*(3): 91-100.

Knowles, R. 1982. "Denitrification." *Microbiol. Rev. 46* : 43-70.

McCarty, P. L., L. Beck, P. St. Amant. 1969. "Biological denitrification of wastewaters by addition of organic materials." In: *Proc. 24th Industrial Waste Conf.*, Lafayette, IN. May 6-8. 1271-1286.

Mercado, A., M. Libhaber, and M. I. M. Soares. 1988. "In situ biological ground water denitrification: concepts and preliminary field tests." *Wat. Sci. Tech. 20*(3): 197-209.

Nuttall, H. E., W. Lutze, L. L. Barton and J. Wolfram. 1997. "In situ bioremediation of contaminated ground water at an UMTRA Site." *International Symposium In Situ and On Site Bioreclamation Proceedings, 1* : 435 – 440.

Nuttall, H. E. 1997. "Biodenitrification of New Mexico Groundwater." *New Mexico Journal of Science, 37*: 64-73

Payne, W. J. 1981. *Denitrification.* Wiley, New York.

DENITRIFICATION KINETICS: MODEL AND COMPARISON TO EXPERIMENTAL DATA

Yongming Lu, H. Eric Nuttall, and Werner Lutze
University of New Mexico, Albuquerque, NM 87131, U.S.A

ABSTRACT: Denitrification kinetics were studied using a mineral medium with denitrifiers collected and enriched from a contaminated groundwater. Kinetic models describe bacterial growth, substrate consumption, and the nitrate to nitrogen gas reduction reactions. The model generated curves that agreed well with experimental data where the initial nitrate concentration ranged as high as 1000 mg/L and the temperature was varied from 12.5 to 24°C. A Arrhenius type expression was used to represent the kinetic model parameters. This approach was successful in allowing the overall kinetic model to incorporate temperature effects. The results showed that the specific growth rate of biomass for either nitrate or nitrite as electron acceptor followed Andrews inhibitory kinetics. In this study using acetate as a substrate, nitrate was reduced sequentially to nitrite, and then to nitrogen gas. We did not observe the nitrogen oxide intermediates between nitrite and nitrogen gas.

INTRODUCTION

Nitrate pollution in groundwater can bring about serious human health problems. It is a widely spread contaminant occurring in many USA States and international locations. In situ biological denitrification is a promising and potentially cost-effective technique for groundwater remediation (Lu, 1998). Fundamental kinetic models require consideration of stepwise reduction of nitrate to nitrogen gas. The competition of nitrate and nitrite for the oxidation of acetate has been investigated using a pure culture of *Pseudomonas fluorescens* (Almeida et al., 1995). Also the dynamic concentrations of nitrate, nitrite, and biomass production were mathematically modeled under conditions of low nitrate concentration using a pure culture of *Pseudomonas denitrificans* (Wang et al., 1995). The purpose of this paper is to develop a fundamental denitrification model that has general application to mixed denitrifying cultures as found in groundwater, and to a wide range of nitrate concentrations, temperatures and pH.

MATERIALS AND METHODS

Denitrifiers were collected from contaminated groundwater at an UMTRA (Uranium Mill Tailings Remedial Action) site near Tuba City, AZ and enriched using an organic medium under anaerobic conditions. The medium contained yeast extract 3 g/L, bacto peptone 5 g/L and KNO_3 2.5 g/L. Acclimation and growth for kinetic study were completed in a mineral liquid medium containing acetate, nitrate and nitrite, in which there were also KH_2PO_4 1.52 g/L, Na_2HPO_4 1.89 g/L, $MgSO_4{\cdot}7H_2O$ 0.2 g/L, and $CaCl_2$ 0.05 g/L. This basal medium was supplemented with 1% v/v of a solution containing $FeSO_4{\cdot}7H_2O$ 0.2 g/L, EDTA

0.5 g/L, and 0.1% v/v of a trace nutrient solution ($ZnSO_4 \cdot 7H_2O$ 0.1 g/L, $MnCl_2 \cdot 4H_2O$ 0.03 g/L, H_3BO_3 0.3 g/L, $CoCl_2 \cdot 6H_2O$ 0.2 g/L, $CuCl_2 \cdot 2H_2O$ 0.01 g/L, $NiCl_2 \cdot 6H_2O$ 0.02g/L, and $NaMoO_4 \cdot 2H_2O$ 0.03 g/L). Amount of added acetate (KCH_3COO), nitrate (KNO_3) and nitrite (KNO_2) were determined by the experimental design. The final pH was 7.0 and the minimum molar ratio of acetate to nitrate for complete denitrification was 6:8 (Lu, 1998). The amount of acetate added was always more than 2 times the amount required in order to assure that the system was not acetate limited.

Kinetic parameter values were evaluated through experiments performed in serum bottles containing 100 ml of autoclaved mineral medium (argon environment) with 0.5 g/L of CH_3COOK. Then based on the particular experimental levels of nitrate and nitrite concentration to be used, 1 to 2 mL of nitrate or nitrite stock solution (10 g/L) and biomass suspension (grown in the organic medium) were added and then the bacteria were incubated at 23±1°C. The intent was for each kinetic run to last long enough for the reaction to be monitored, but also to be completed in just a few hours. These prepared serum bottles were then used in the kinetic runs for evaluating kinetic parameter values. Each bottle was injected with an amount of nitrate or nitrite stock solution so that each run was performed at a different initial NO_3^- or NO_2^- concentration in the range of 25 to 250 mg/L. During each kinetic run, liquid was sampled frequently and analyzed for biomass and ion concentrations. Experiments were performed in the temperature range of 12.5~24°C by placing the serum bottles in a temperature controlled water bath unit. Biomass concentration was monitored by measuring the optical density (turbidity) of liquid samples at a wavelength of 500 nm through a calibration procedure. Concentrations of acetate, nitrate and nitrite were analyzed through ion chromatographic (IC) analysis of liquid samples. Samples were diluted and analyzed immediately after filtration for biomass removal. Experimentally determined parameter values were then used in the mathematical models to simulate complete denitrification curves and compare results to experimental data from separate runs.

MODELING THEORY

In order to model the denitrification process, two sets of experiments were conducted to evaluate yield coefficients, specific degradation rates, and kinetic parameters in the specific growth rate expressions for **denitratation**, $NO_3^- \rightarrow NO_2^-$, and **denitritation**, $NO_2^- \rightarrow N_2(g)$.

Parameter Evaluation. Under constant temperature and pH, bacterial growth is limited by the availability of substrates. Because of endogenous metabolism (in terms of specific degradation rate, μ_e), true specific growth rate (μ) is not equal to observed specific growth rate (μ_{net}), and true yield (Y_{xs}) is not equal to apparent yield (Y_{XS}^{app}) either. Logarithmic growth data from runs with different initial substrate concentrations were used to generate pairs of (Y_{XS}^{app}, μ_{net}) values. The values of Y_{xs} and μ_e were graphically determined using Eq. 1 to fit the data. The relation between limiting substrate (nitrate S_1 or nitrite S_2) and specific growth rate ($\mu=\mu_{net} + \mu_e$) may be expressed as Andrews model as presented in Eq. 2 (Andrews, 1968), when substrate inhibits growth at high concentration.

$$\frac{1}{Y_{XS}^{app}} = \frac{1}{Y_{XS}} + \frac{\mu_e}{Y_{XS}} \frac{1}{\mu_{net}} \tag{1}$$

$$\mu_{net} = \frac{\bar{\mu}_{\max} S}{K_S + S + S^2 / K_i} - \mu_e \tag{2}$$

Kinetics Prediction. Eqs. 3 to 5 are used to describe the concentration change of biomass(X), nitrate(S_1) and nitrite(S_2) during denitrification, $NO_3^- \rightarrow NO_2^- \rightarrow N_2(g)$. Subscripts of 1 and 2 represent parameters related to nitrate and nitrite respectively. Inclusion of step functions in the models ($\Delta(S)=1$ if $S>0$, and $\Delta(S)=0$ if $S>0$) implies that specific degradation rate associated with a substrate is zero when that substrate is not present in the medium. The constant α in Eq. 5 is defined as the amount of nitrite produced based on unit amount of nitrate consumed (mg mg^{-1}). α is calculated using Eq. 6 based on mass balance of nitrogen atoms. M is the molecular weight of biomass and σ is the number of nitrogen atoms in the molecular formula of biomass. M = 113 and σ =1 if the molecular formula for biomass is assumed to be $C_5H_7O_2N$. The specific growth rate expression for biomass shown in Eq. 3. The rate equations of both nitrate and nitrite substrate are shown in Eqs 4 and 5. We used a modified Andrews model to describe the specific growth rate of denitratation, $\mu_1(S_1, S_2)$, and denitritation, $\mu_2(S_2, S_1)$, in which cross-inhibition between nitrate and nitrite is considered.

$$\frac{dX}{dt} = [\, \mu_1(S_1, S_2) - \mu_{e,1}\, \Delta(S_1) \; + \; \mu_2(S_2, S_1) - \mu_{e,2}\, \Delta(S_2)]\, X \tag{3}$$

$$\frac{dS_1}{dt} = -\frac{1}{Y_{XS_1}} \mu_1(S_1, S_2)\, X \tag{4}$$

$$\frac{dS_2}{dt} = \frac{\alpha}{Y_{XS_1}} \mu_1(S_1, S_2)\, X - \frac{1}{Y_{XS_2}} \mu_2(S_2, S_1)\, X \tag{5}$$

$$\alpha \; = \; 46\,(\frac{1}{62} - \frac{Y_{XS_1}\, \sigma}{M}) \tag{6}$$

$$\mu_1(S_1, S_2) = \frac{\bar{\mu}_{\max,1}\, S_1}{K_{S,1} + S_1 + S_1^2 / K_{i,1} + K_{21}\, S_2\, S_1} \tag{7}$$

$$\mu_2(S_2, S_1) = \frac{\bar{\mu}_{\max,2}\, S_2}{K_{S,2} + S_2 + S_2^2 / K_{i,2} + K_{12}\, S_1\, S_2} \tag{8}$$

The concentration of acetate during denitrification can be determined in terms of nitrate, nitrite and biomass concentrations based on a mass balance of nitrogen (Lu, 1998). Acetate is consumed for both bacterial respiration (Eqs. 9 and 11) and synthesis (Eqs. 10 and 12) for both denitratation (Eqs. 9 and 10) and denitritation (Eqs. 11 and 12).

$$CH_3COO^- + 4NO_3^- \rightarrow 4NO_2^- + 2HCO_3^- + H^+ \tag{9}$$

$$7\,CH_3COO^- + 2\,NO_3^- + 5\,H^+ \rightarrow 2\,C_5H_7O_2N + 4\,HCO_3^- + 4\,H_2O \qquad (10)$$

$$3\,CH_3COO^- + 8\,NO_2^- + 5\,H^+ \rightarrow 4\,N_2 + 6\,HCO_3^- + 4\,H_2O \qquad (11)$$

$$13CH_3COO^- + 4NO_2^- + 11H^+ \rightarrow 4C_5H_7O_2N + 6HCO_3^- + 8H_2O \qquad (12)$$

Effect of Temperature. An Arrhenius expression is used to represent temperature dependence of all parameters in the kinetic models. P is a general symbol representing any of the following parameters: $\mu_{e,1}$, $\mu_{e,2}$, $\bar{\mu}_{max,1}$, $\bar{\mu}_{max,2}$, $K_{S,1}$, $K_{S,2}$, $K_{i,1}$, $K_{i,2}$. K_{12}, K_{21}, Y_{XS_1}, and Y_{XS_2}. F is the frequency factor, Ea is activation energy, T is absolute temperature, and R is universal gas constant.

$$P = F\ Exp\,(-\,E_a\,/\,RT) \qquad (13)$$

RESULTS AND DISCUSSION

Experiments were performed at 12.5 to 24°C and pH=7.0 due to the groundwater strongly buffered by the presence of minerals. To calculate μ_{net}, the biomass concentration (dry basis) for each experimental run was plotted on a semi-logarithmic graph versus time and substrate (nitrate or nitrite) concentration data from the same run were plotted against each other in order to calculate Y_{XS}^{app}. Initial slopes were used in both cases. Pairs of (μ_{net}, S) and (Y_{XS}^{app}, μ_{net}) values were generated from the experiments runs at different initial nitrate or nitrite concentrations. Pairs of (Y_{xs}^{app}, μ_{net}) were used to determine Y_{XS} and μ_e based on Eq. 1. Strong maintenance requirement was observed and can be described as self-oxidation of portions of the biomass. After specific degradation rates were determined, the net specific growth rates were plotted versus corresponding substrate concentrations. The net specific growth rate data were regressed to the expression of μ_j - $\mu_{e,\,j}$ (Eq. 2, j=1: nitrate, j=2: nitrite, with $\mu_{e,\,j}$ held constant) in order to calculate values of $\bar{\mu}_{max,\,j}$, $K_{S,\,j}$, and $K_{i,\,j}$. Fig. 1. Shows an example. It is clear from these plots that, after an initial increase, the net specific growth rates drop with increasing concentration of both nitrate and nitrite. This phenomenon shows the limitation of high substrate concentration on the bacterial growth. The values of K_{12} and K_{21} were determined by fitting the experimental data using Eqs 3 to 5. The parameter values obtained are shown in Table 1. All parameters except Y_{XS_1}, Y_{XS_2}, K_{12}, and K_{21} are found to be temperature dependent. The temperature dependency was successfully described using an Arrhenius expression, and Ea and F values generated are shown in Table 2. It was found that $K_{12} > 5$ L/mg and $K_{21} < 0.001$ L/mg gave satisfactory fitting to all the experimental runs at temperatures ranged from 12.5 to 24°C. K_{12}=10 and K_{21}=0 were used in modeling calculation, as nitrate was always reduced sequentially to nitrite and then to nitrogen gas. In other words, nitrite reduction was inhibited completely in the presence of nitrate. High concentrations of nitrite had little inhibition on nitrate reduction, when acetate was used as an electron acceptor and carbon source.

TABLE 1A. Summary of parameter values used for denitrification process prediction at pH=7.0 (Parameters for denitratation).

T (°C)	$\bar{\mu}_{max,1}$ (hr^{-1})	$K_{S,1}$ (mg/L)	$K_{i,1}$ (mg/L)	$\mu_{e,1}$ (hr^{-1})
12.5	0.2199	71.587	345.950	0.03843
16	0.2962	65.998	372.183	0.04990
20	0.3558	41.208	457.742	0.05753
24	0.3787	33.914	508.874	0.06485

TABLE 1B. Summary of parameter values used for denitrification process prediction at pH=7.0 (Parameters for denitritation).

T (°C)	$\bar{\mu}_{max,2}$ (hr^{-1})	$K_{S,2}$ (mg/L)	$K_{i,2}$ (mg/L)	$\mu_{e,2}$ (hr^{-1})
12.5	0.05852	86.940	169.880	0.00774
16	0.08424	59.150	209.394	0.0109
20	0.13079	43.966	271.962	0.0215
24	0.19070	27.218	336.549	0.0270

TABLE 1C. Summary of parameter values used for denitrification process prediction at pH=7.0 (Parameters which are independent of temperature).

α (mg/mg)	Y_{XS_1} (mg/mg)	Y_{XS_2} (mg/mg)	K_{12} (L/mg)	K_{21} (L/mg)
0.6172	0.306	0.300	10	0

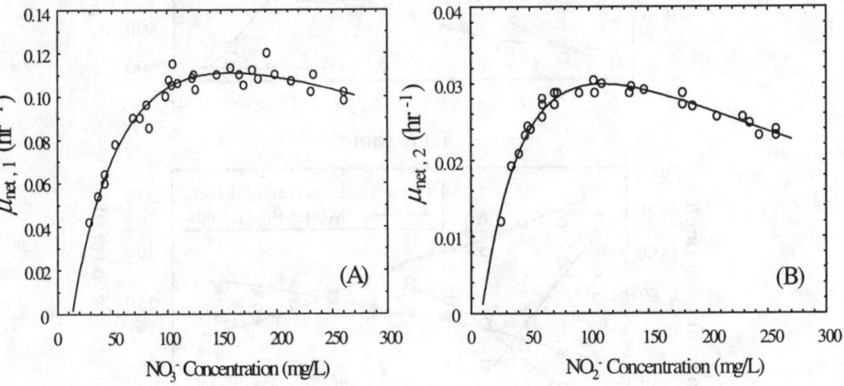

FIGURE 1. Net specific growth rate curves for the denitrifying bacterial culture at 16°C and pH=7.0 on nitrate (Case A) and nitrite (Case B).

Fig. 2 shows the results the model predictions versus experimental data for two examples: one at 16°C and the other at 24°C . Note that nitrite begins to reduce only after it reaches its local maximum and all the nitrate is depleted. This

TABLE 2A. Arrhenius constants for the kinetic parameters in the denitratation model (pH=7.0).

Constants	$\bar{\mu}_{max,1}$ (hr^{-1})	$K_{S,1}^{-1}$ (L/mg)	$K_{i,1}$ (mg/L)	$\mu_{e,1}$ (hr^{-1})
F*	$2.608*10^5$	$1.779*10^7$	$1.310*10^7$	$2.070*10^4$
Ea (kJ/mole)	33.062	49.923	25.075	31.232

• Units are the same as those for the corresponding kinetic parameter

TABLE 2B. Arrhenius constants for the kinetic parameters in the denitritation model (pH=7.0).

Constants	$\bar{\mu}_{max,2}$	$K_{S,2}^{-1}$ (L/mg)	$K_{i,2}$ (mg/L)	$\mu_{e,2}$ (hr^{-1})
F*	$1.315*10^{12}$	$5.294*10^{10}$	$9.553*10^9$	$5.499*10^{12}$
Ea (kJ/mole)	73.011	69.246	42.380	81.213

*Units are the same as those for the corresponding kinetic parameter

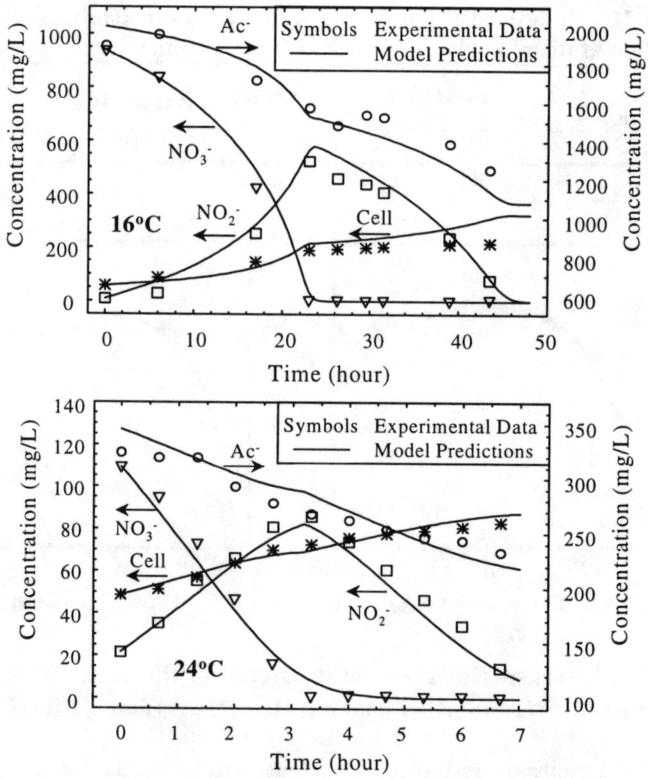

FIGURE 2. Comparison of experimental data (symbols) to model prediction (solid lines) for two batch experiments at 16°C or 24°C at pH=7.0.

sequential reduction phenomenon can be explained by competitive inhibition of nitrite reduction by nitrate. The agreement of the model prediction curves with the experimental data is reasonably good. Consequently, the models and parameter values presented here are applicable for a wide range of substrate concentrations and temperatures.

REFERENCES

Almeida, J. S., M. A. M. Reis, and M. J. T. Carrondo. 1995. "Competition Between Nitrate and Nitrite Reduction in Denitrification by *Pseudomonas fluorescens*." *Biotechnol. Bioeng. 46*:476-484.

Andrews, J.F. 1968. "A Mathematical Model for the Continuous Culture of Microorganisms Utilizing Inhibitory Substrates." *Biotechnol. Bioeng. 10*:707-723.

Lu, Y. 1998. *Sequential Bioremediation of Nitrate and Uranium in Contaminated Groundwater*, Ph.D. Dissertation, University of New Mexico, NM, USA.

Wang, J-H., B. C. Baltzis, and G. A. Lewandowski. 1995. "Fundamental Denitrification Kinetic Studies with *Pseudomonas denitrificans*." *Biotechnol. Bioeng. 47*:26-41.

IN SITU AUTOTROPHIC DENITRIFICATION USING ELECTRODE UNDER OLIGOTROPHIC CONDITIONS

Yutaka Sakakibara (Gunma University, Gunma, Japan)
Jun Kusaka (Graduate School of Gunma University, Gunma, Japan)

ABSTRACT: The feasibility of an *in situ* autotrophic denitrification process was investigated under oligotrophic conditions. Electrodes were embedded in laboratory-scale sandy aquifers and H_2 gas was produced through the electrolysis of water for enhancing autotrophic denitrification. In experiment, a groundwater containing about 20 mg-N/L of $NaNO_3$ and no electron donor was fed continuously, while trace nutrients such as phosphate were not injected. Experimental results showed that, when carbon electrodes were embedded in a direction perpendicular to groundwater flow, the complete and stable denitrification was achieved over one year without clogging problem. Denitrification rate varied depending on applied current. Electric power requirement was in the range of 0.03~0.09 kWh/g-N. This energy requirement corresponded to about 8 % of total energy obtained from a wind or solar power generation at some places in Japan.

INTRODUCTION

An increase in nitrate concentration is becoming a problem in many sources of groundwater supply. *In situ* denitrification is regarded as an economical process in comparison with other on-site processes such as ion exchange and reverse osmosis. Although extensive studies have been conducted in Eastern Europe (Mateju et al., 1992), most studies focused on heterotrophic denitrification in which organic substrate and trace nutrients were injected through wells. Several problems, such as clogging of aquifer caused by excessive growth of indigenous heterotrophs and pollution by injected substrates, were sometimes encountered.

Autotrophic, H_2-utilizing denitrification is a relatively new technology and is receiving attention for groundwater purification. Sakakibara et al. (1995, 1997) proposed a combined process of electrolytic H_2 production with autotrophic denitrification and showed its usefulness and simplicity in operation. In the autotrophic process, the injection or addition of trace nutrients such as phosphate may not be required because of smaller growth yield compared to that of heterotrophs. In addition, the use of electrode may enable the process to utilize natural energy by installing a wind or solar power generation device.

This study examined the stability of *in situ* autotrophic denitrification under oligotrophic conditions. Effects of electrode configuration and electric currents on denitrification performance were also examined as well as electric power requirements and the possibility of utilizing natural energy.

Objective. The primal objective of this study is to know the feasibility of a novel *in situ* autotrophic denitrification process from laboratory-scale experiments. The experimental goal is to achieve stable denitrification over one year under oligotrophic condition by an application of electric currents.

MATERIALS AND METHODS

Experimental Setup. Laboratory-scale, experimental apparatus used in this study is illustrated in Figure 1. Sandy aquifer is built in a rectangular plastic housing whose effective volume is about 10 L. Solution containing about 20 mg-N/L of $NaNO_3$ and no electron donor was fed continuously by a peristaltic pump. To realize uniform water flow, porous plates were set at the inlet and outlet of the aquifer. After seeding facultative anaerobes, electric currents were applied and groundwater quality in aquifer was analyzed. In experiment, trace nutrients such as phosphate were not added in the feed solution, but it originally contained about 20 μ g/L of TP.

Electrode. Carbon electrodes and expanded and rod-type metal electrodes were used in this study. These electrodes were embedded in aquifer in two configurations; parallel (PL) and perpendicular (PP) configurations. In those configurations, a pair of cathode and anode was set in the directions parallel and perpendicular to groundwater flow, respectively. Four experimental setups were illustrated in Figure 1. In the parallel configuration, cathode was placed at upflow position to anode.

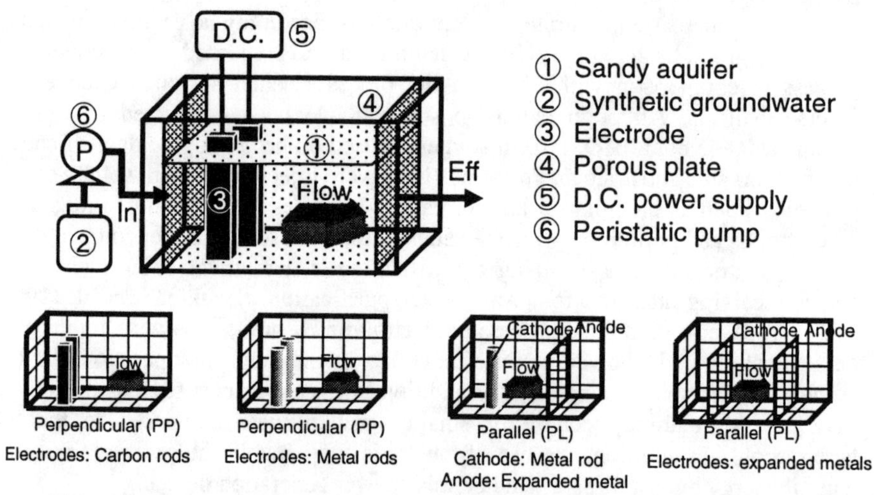

Figure 1. Experimental apparatus and configuration of electrodes.

Sampling and Analysis/Measurement. Water samples were taken out from aquifer and measurements were made for nitrate, nitrite, pH, and ORP. Nitrate and nitrite were analyzed using Ion Chromatography (IC7000, Yokogawa Analytical Systems) as well as a Spectrometer (Shimadze, UV-150) after the coloration by Brucine method. Distribution of electric potential (ΔE) based on the cathode was measured using multi-meter (Yokogawa, Model 7532). Details of the experimental methods were shown elsewhere (Sakakibara et al., 1995).

RESULTS AND DISCUSSION

Nitrate Removal. Figure 2 shows comparisons of nitrate concentration in the effluent for the parallel (PL) and perpendicular (PP) configurations. In the parallel (PL) configuration of rod and expanded metal electrodes or two expanded metal electrodes, the effluent concentration was almost the same as the influent. On the other hand, the effluent nitrate concentration declined with electric current, when using carbon electrode as anode in the perpendicular configuration. Nitrite accumulation and clogging problem were not observed during the experiment. According to Kurt et al. (1987), autotrophic, H_2-utilizing denitrification is represented by Reaction (1); whereas H_2 production through the electrolysis of water is by Reaction (2).

$$2NO_3^- + 5H_2 = N_2 + 4H_2O + 2OH^- \quad (1)$$
$$2H_2O + 2e^- = H_2 + 2OH^- \quad (2)$$

In the perpendicular configuration of carbon electrodes, pH was kept around neutral regardless of the extent of denitrification. This is attributed to the following anodic reaction (Kinoshita, 1988); where the amount of produced CO_2 is suitable for neutralizing alkalinity produced in denitrification process (Sakakibara et al, 1997).

$$C + 2H_2O = CO_2 + 4H^+ + 4e^- \ (e_0{=}0.2\text{~}0.5V) \quad (3)$$

When using metal anode in the perpendicular configuration, extensive denitrification did not occur under the experimental condition in this study. This is due to inhibitory effect by oxygen produced from cathode. A calculation result based on the conventional mass transfer theory, however, showed that denitrification may occur when a sufficient electrode distance exists between the anode and cathode.

Based on Reactions (1) and (2), current-denitrification efficiency was estimated to be 0.6 (-) in the perpendicular configuration of carbon electrodes.

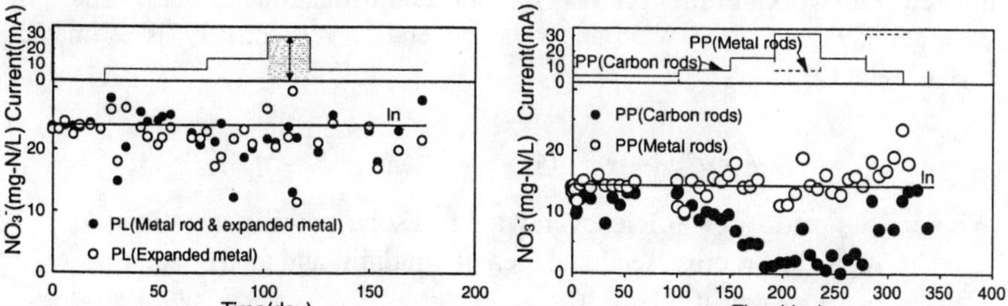

Fig 2. Comparisons of the effluent nitrate concentrations in parallel (PL) and perpendicular (PP) configurations.

Effects of Electric Field. Figure 3 shows distributions of electric potential (ΔE) and pH profiles in aquifer. In the perpendicular configuration, a steep decline in potential drop was observed between the anode and cathode, whereas there existed a plane region downstream the electrodes. In contrast to the perpendicular configuration, a steeper gradient of electric potential was built in aquifer downstream the cathode. This potential gradient caused a polarization and thereby created an increase and decrease in pH at places near cathode and anode, respectively. As shown in Figure 3(c), it was thought that relatively high pH values above 10 doesn't suit for the enhanced denitrification.

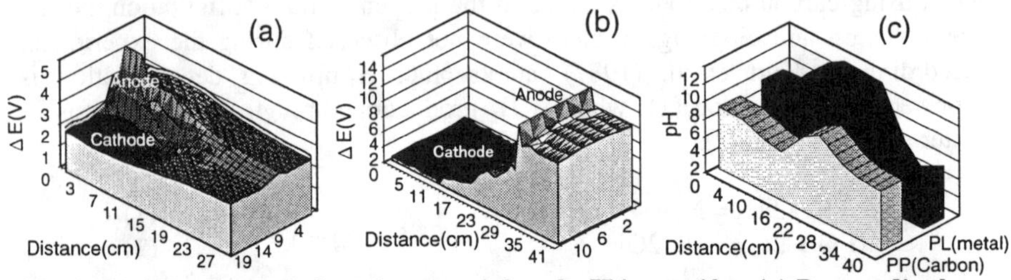

Fig 3. Distributions of electric potential and pH in aquifer. (a) Perpendicular configuration. (b) Parallel configuration. (c) Comparison of pH profiles for perpendicular and parallel configurations.

From Figures 3(a)~(c), it is shown that the perpendicular electrode configuration was better than the parallel configuration due to smaller effect of polarization reflected by electric field. In addition, the use of carbon electrode has advantages for both creating anoxic condition and neutralizing water. A smaller electrode potential compared to that for O_2 evolution (e^0=1.23 V) is also advantageous from the economical point of view. Electric power requirement to reduce one gram of nitrate-N was in the range from 0.03 to 0.09 kWh/g-N, while electrode potentials ranged from 2 to 4 V.

Possibility of Utilizing Natural Energy. Recent development of technology has been increasing the cost performance significantly for wind and solar power generations (Danish Wind Turbine Manufactures Association, 1998; Kuki, 1996). It is thought renewable energy such as wind power or solar radiation can be utilized for working the present *in situ* denitrification process. The power generation, P_w and P_s in wind and solar generation respectively, is estimated by Eqs. (4) and (5) :

$$P_w = \frac{1}{2}\eta_w A_w \rho u^2 \quad (4), \qquad P_s = \eta_s A_s S \quad (5)$$

Where η_w and η_s = efficiency in wind and solar power generation (-)
A_w and A_s = cross sectional area of windmill and area of solar panel (m^2)
ρ_w = air density (kg/m^3)
S = solar radiation (kW/m^2)

Figure 4 shows a result of electric power generation estimated from local weather data (Japan Weather Association, 1997) at 3 places in Japan, i.e. Ishigakizima in Okinawa, Wakkanai in Hokkaidou, and Maebashi in Gunma, respectively. When installing 150 to 600 kW Wind Turbines (Danish Wind Turbine Manufactures Association, 1998), annually available energies were calculated to be in the range from 50 to 600 MWh. In Ishigakizima and Wakkanai, an energy of 400 MWh was estimated for 300 kW wind turbine. Larger wind turbines could produce larger electric energy, but efficiency for generation was higher in smaller turbine. In cases of solar power generation having the same panel area as that of wind turbine, annually available energy was 100~400 MWh, in which η_s was assumed to be 0.172 (Shin Energy Binran, 1995).

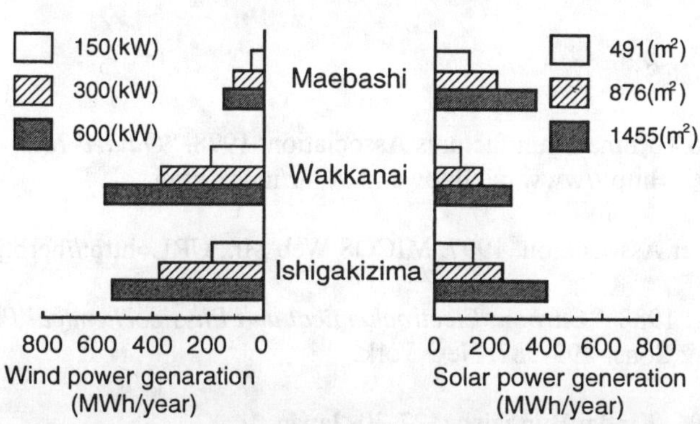

Fig 4. Wind and solar power generations estimated from local weather data.

Table 1. Electric energy consumption in a model village

Number of household	100 households
Electric energy consumption per household	280kWh/month
Electric energy consumption in a model village	300MWh/year
Obtainable electric energy from natural energy	360MWh/year
Electric energy consumption for nitrate removal	30MWh/year
Electric energy consumption for public facilities	30MWh/year
Flow rate of groundwater	1m/day~1m/year
Per capita water consumption	$0.3m^3/h$ /day
Water consumption of a model village	$100m^3/day$

Table 1 lists an example of electric energy consumption in a model village comprised of 100 households, where a renewable energy of 400 MWh is being produced from a wind or solar power generation. Annual water consumption and household electric power consumption are estimated to be 0.37 Mm^3 and 300 MWh, respectively (TEPCO, 1977). If the village depends on groundwater and 10 mg-N/L of nitrate should be removed, electric energy of about 30 MWh/year will be required. This energy consumption corresponds to 8 % of total power generation. This figure is not regarded as unrealistic, but further study will be needed to reduce the energy consumption.

CONCLUSIONS

Stable denitrification performance was demonstrated over one year under oligotrophic (no addition of trace nutrients) conditions in the present *in situ* process. The use of carbon electrode and its configuration perpendicular to groundwater flow were effective for enhanced and stable denitrification. Electric power requirement was in the range of 0.03 to 0.09 kWh/g-N removed, which was considered obtainable from natural energy.

ACKNOWLEDGMENT

This study is partially supported by the grant-in-aid for the Development of Scientific Research (#08650639) and the River Environment Fund (Project #10-1-⑦-2).

REFERENCES

Danish Wind Turbine Manufactures Association. 1998. "*Guided Tour on Wind Energy.*" URL=http://www.windpower.dk/tour/index.html.

Japan Weather Association. 1997. MICOS Web Mr., URL=http://herb.jwa.go.jp/.

Kinoshita, K. 1988. "*Carbon, Electrochemical and Physicochemical Properties.*" John Wiley & Sons: 293-387, New York.

Kuki, K. 1996. Kindai Bungeisha: 27-30, Japan.

Kurt, M., I. J. Dunn, J. R. Bourne. 1987. "*Biological Denitrification of Drinking Water Using Autotrophic Organisms with H_2 in a Fluidized-Bed Biofilm Reactor.*" Biotechnol. Bioeng. **29**: 493-501.

Mateju, V., S. Cizinska, J. Krejci, and T. Janoch. 1992. "*Biological Water Denitrification - A Review.*" Enzyme Microb. Technol. **14**: 170-183.

Sakakibara, Y., T. Tanaka, K. Ihara, T. Watanabe, and M. Kuroda. 1995. "An In-situ Denitrification of Nitrate-Contaminated Groundwater Using Electrodes." *Proc. of Environmental Engineering Research* **32**:407-415.

Sakakibara, Y., K. Araki, T. Watanabe, and M. Kuroda. 1997. "*The Denitrification and Neutralization Performance of an Electrochemically Activated Biofilm Reactor Used to Treat Nitrate-Contaminated Groundwater.*" Wat. Sci. Tech. **36**(1): 61-68.

Shin Energy Binran 1995. Shigen Energy tyou, Tusyo Sangyou Tyousakai.

TEPCO Data Base. 1997. URL=http://www.tepco.co.jp/corp-com/db/.

COMBINED MICROBIAL-Fe(0) SYSTEM TO TREAT NITRATE-CONTAMINATED GROUNDWATER

Todd D. DeJournett and Pedro J.J. Alvarez
(The University of Iowa, Civil & Environmental Engineering, Iowa City, Iowa)

ABSTRACT: Experiments were conducted to delineate the applicability and limitations of biologically active Fe(0) barriers to remove nitrate under various geochemical and hydraulic conditions. Microcosm studies showed that, while no Fe(0) treatment was needed to remove nitrate from high-carbon soils, adding Fe(0) to low-carbon soil can supplement the electron donor pool and enhance nitrate removal. Montmorillonite, an acidic aluminosilicate mineral, enhanced Fe(0) corrosion and nitrate removal, and reduced the transient accumulation of nitrite. Combining autotrophic denitrifiers (e.g., *Paracoccus denitrificans*) with Fe(0) significantly reduced the amount of nitrite eluted from aquifer columns. Bacteria were observed to preferentially colonize the Fe(0) surface, which produces cathodic H_2 when corroded by water. The preferential colonization of Fe(0) suggest that hydrogenotrophic consortia are likely to develop around Fe(0) walls to exploit cathodic depolarization as a metabolic niche.

INTRODUCTION

Modern agricultural practices and increased use of nitrogenous fertilizers have resulted in increased amounts of nitrate entering groundwater systems (Kapoor and Viraraghavan, 1997). Nitrate in drinking water is of concern because it poses adverse human health risks. It can be reduced to nitrite in saliva and the gastrointestinal tract, which causes methemoglobinemia (blue baby syndrome). Methemoglobinemia primarily occurs in infants, but may also occur in adults deficient in glucose-phosphate dihydrogenase (Challis, 1973). To prevent this, a primary drinking water standard of 10 mg/L $_{NO3}^{-}$-N has been set in the US.

Although numerous physical-chemical processes have been proposed for removal of nitrate from water, many of these have proven marginally cost-effective and/or have detrimental side-effects on water quality. The use of autotrophic, hydrogen-utilizing denitrifiers is an attractive technique, due primarily to the cleanliness of the process. Since CO_2 is the carbon source for autotrophic growth, there is no need to remove residual organic substrates. H_2 gas is inherently clean and only slightly soluble in water (0.78 mg/L at 1 atm, 25°C). Autotrophic growth also leads to lower cell yield and less biomass production. Nevertheless, the relatively high cost and low solubility of hydrogen represents a limitation for delivering this energy source to autotrophic denitrifying bacteria.

Research with hydrogenotrophic, anaerobic bacteria suggests a method by which iron corrosion could be exploited to overcome limitations associated with hydrogen delivery in autotrophic denitrifying systems. Daniels *et al.* (1987) demonstrated that when zero-valent iron (Fe(0)) is submerged in anoxic water, hydrogen gas is produced via cathodic depolarization:

$$Fe^0 + 2H_2O \Leftrightarrow Fe^{2+} + H_2 + 2OH^- \quad (1)$$

The use of this cathodic hydrogen as an energy source to support bacterial growth has been demonstrated for several types of anaerobic pure cultures, including autotrophic denitrifiers (Till *et al.,* 1998). In such cases, nitrate is eventually reduced via nitrite to N_2 gas. Abiotic reduction of nitrate by Fe(0) has also been studied (Huang *et al.,* 1998; Till *et al.,* 1998), and proceeds through a different pathway leading to the less desirable end product, NH_4^+.

This work investigated the applicability and limitations of combining hydrogenotrophic denitrifiers with Fe(0) to remove nitrate under various geochemical and hydraulic conditions. Variables considered included (1) the organic carbon content of the aquifer material, which represents a potential electron donor pool to sustain denitrification; (2) the presence of aluminosilicate minerals, which may buffer the Fe(0)-corrosion-induced increase in pH that inhibits microorganisms; and (3) the limits of hydraulic loadings that can be handled by a biologically-active Fe(0) barrier to intercept and treat nitrate plumes.

MATERIALS AND METHODS

Aquifer microcosms were prepared in duplicate using 250 mL serum bottles capped with Mininert valves. All microcosms were amended with 90 grams of either 2.5% org-C soil or 0.1% org-C sand plus 100 mL of mineral medium containing 30 mg/L NO_3-N. Some microcosms were amended with montmorillonite clay. These contained 80 grams of low-carbon soil and 10 grams of montmorillonite. Biological and abiotic transformations were studied separately and interactively by including or excluding Fe (0) (10g Fisher Fe-filings) and denitrifying hydrogenotrophs (10 mL of *Paracoccus denitrificans*, ATTC# 17741 suspension, OD_{600}=0.006). Abiotic removal was assessed per NH_4^+ accumulation, and biological denitrification was quantified per N_2O accumulation in acetylene-blocked microcosms (Yoshinari and Knowles, 1976).

The feasibility of bioaugmenting Fe(0) in sandy soil was also investigated in continuous-flow columns. Glass columns were packed with a mixture of sand and Fe^0 filings (15% by weight). One column was autoclaved to discourage growth of indigenous microbes, while the other was seeded with a liquid culture of *P. denitrificans* (OD_{600}=10). Mineral medium containing 33 mg/L NO_3^--N was pumped through the columns in an upflow mode using a Masterflex® (Barrington, IL) 7523-30 peristaltic pump with a 519-15 pump head. The flowrate was varied from 0.02 mL/min to 0.16 mL/min (i.e., HRT = 2 to 0.25 days). The influent reservoirs were 5000 mL Pyrex jars topped with Teflon®-lined delivery caps. These reservoirs were continuously purged with an N_2/CO_2 (80:20 v/v) gas mixture using tubing with a stone diffuser. At the end of the 152 day study, samples were taken from the soil/Fe(0) matrix and examined under a scanning electron microscope.

All anions were measured using a DioneX BioLC ion chromatograph. Ammonium was measured using a Dionex DX-100 ion chromatograph.

RESULTS AND DISCUSSION

Semi-permeable reactive Fe(0) barriers are particularly attractive for groundwater remediation in that they conserve energy and water, and through long-term low operating and maintenance costs, have the potential to be considerably less costly than conventional cleanup methods.. Early studies of Fe(0) barriers found little microbial contribution to contaminant degradation, and research has focused primarily on abiotic processes. Nevertheless, the potential for microorganisms to enhance reductive treatment with Fe(0) has been demonstrated in recent laboratory experiments (Till *et al.*, 1998; Weathers *et al.*, 1997). This work corroborates that microbial utilization of cathodic hydrogen is a critical link between biogeochemical interactions and enhanced contaminant removal.

Batch Aquifer Microcosms. Microcosms prepared with high-carbon (2.5%) top soil removed nitrate relatively fast (ca. 30 mg/L as N within 4 days), regardless of whether they were prepared without amendments or amended with Fe(0), *P. denitrificans*, or both. Apparently, indigenous microorganisms utilized naturally occurring organic matter as electron donors to denitrify. Thus, neither Fe(0) nor bacterial amendment enhanced NO_3^- removal in high-carbon top soil.

In contrast, the addition of Fe(0) to low-carbon (0.1%) sandy soil had a beneficial effect, resulting in the removal of nitrate below the MCL (10 mg/L as N) within 20 days (Figure 1). The no-treatment controls exhibited no nitrate removal, possibly due to a low concentration of suitable electron donors in the sandy soil. Adding Fe(0) supplemented the electron donor pool and enhanced nitrate removal. Adding *P. denitrificans* to Fe(0)-amended microcosms did not enhance nitrate removal. Fe(0) corrosion increased the pH in these microcosms to a level that inhibited denitrifiers (pH$\geq$10.0 after 25 days). This notion is supported by the absence of N_2O in acetylene-blocked replicates.

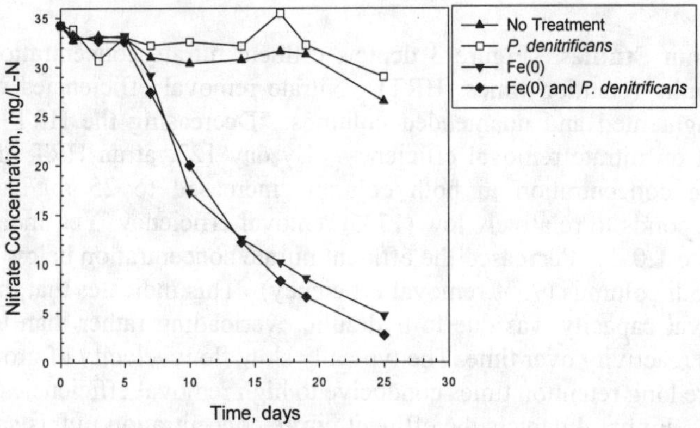

FIGURE 1: Nitrate removal in low-carbon aquifer microcosms. Fe(0) enhanced nitrate removal to a greater extent than bioaugmentation, possibly due to a corrosion-induced increase in pH that inhibited bacteria.

Montmorillonite clay has been reported to buffer against the corrosion-induced pH increase in Fe(0) systems. This is accomplished through a series of dissolution and proton-generating reactions at the Fe(0) surface (Powell and Puls, 1997). In order to provide a more hospitable environment for microbial growth, montmorillonite was used as a buffer against pH increase. Figure 2 depicts the ability of montmorillonite to enhance nitrate removal by Fe(0) in soil microcosms. Montmorillonite also prevented the accumulation of the more toxic intermediate, nitrite. These effects were attributed to enhanced Fe(0) corrosion at a lower pH. Enhanced Fe(0) reactivity in montmorillonite-amended microcosms was also indicated by higher H_2 evolution. Thus, the presence of naturally occurring aluminosilicate minerals may improve the performance of Fe(0) barriers.

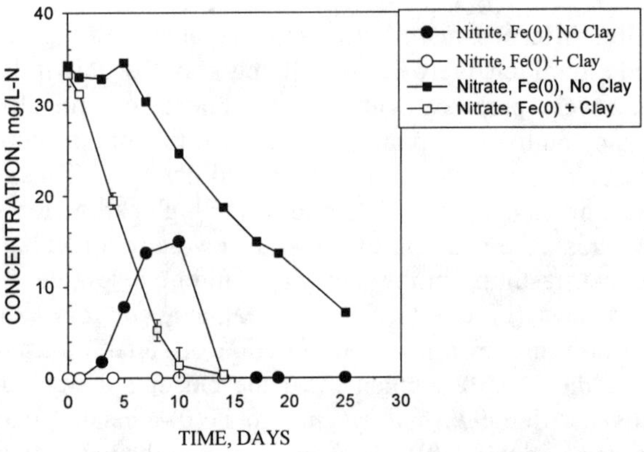

FIGURE 2: Effect of montmorillonite clay (11% by weight) on abiotic nitrate removal by Fe(0) and transient accumulation of nitrite.

Column Studies. Figure 3 depicts effluent nitrate concentrations for different hydraulic retention times (HRT). Nitrate removal efficiencies were similar for bioaugmented and unamended columns. Decreasing the HRT had an adverse effect on nitrate removal efficiency. By day 123, at an HRT of 0.25 days, the nitrate concentration in both columns increased to 25 mg/L NO_3^--N. This corresponds to relatively low (17%) removal efficiency. Yet, increasing the HRT back to 1.0 day decreased the effluent nitrate concentration below 10 mg/L (as N) for both columns (73% removal efficiency). This indicates that the loss of nitrate removal capacity was due to hydraulic overloading rather than to exhaustion of Fe(0) reactivity over time. The typically slow flow velocity of groundwater would ensure long retention times conducive to high removal efficiencies.

Figure 4 depicts the effluent nitrite concentration in this same experiment. Nitrite concentrations in the effluent of the unamended column increased to as high as 7 mg/L NO_2^--N early in the study, while nitrite concentrations in the effluent of the bioaugmented column remained below the MCL (1 mg/L NO_2^--N). This was attributed to microbial denitrification of nitrite to N_2. As the columns acclimated over time, nitrite was eliminated from the effluent of both columns.

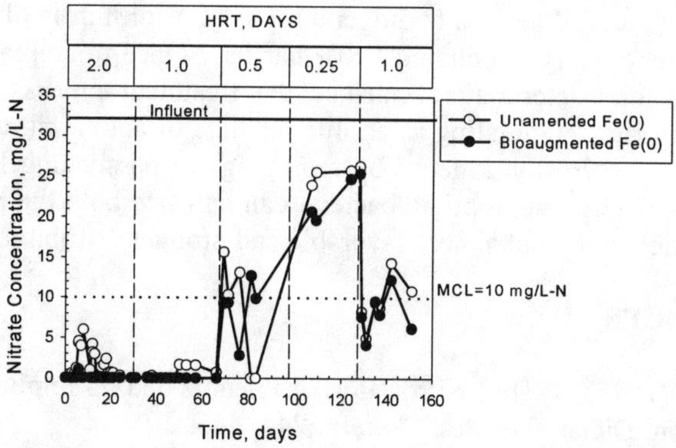

FIGURE 3. Effect of hydraulic retention time (HRT) on the effluent nitrate concentration from aquifer columns.

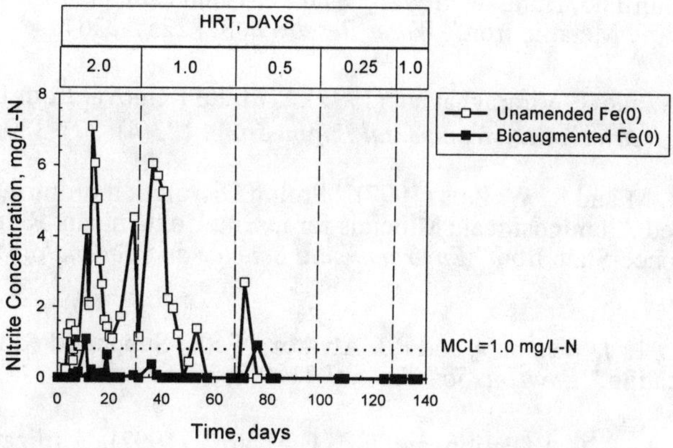

FIGURE 4. Effluent nitrite concentration from continuous flow columns.

The soil and Fe(0) filings from the columns were analyzed at the end of the experiment at the UI Central Microscopy Research Facility using scanning electron microscopy. Bacteria were found to selectively colonize the Fe(0) surfaces in both columns. This suggests that indigenous denitrifiers are likely to colonize Fe(0) barriers to exploit cathodic depolarization as a metabolic niche. The selective colonization of Fe (0) surfaces may facilitate microbial uptake of relatively insoluble cathodic H_2.

In summary, biologically-active Fe(0) barriers hold great promise as a cost-effective alternative for treating groundwater contaminated with nitrate, and should also be considered to treat other redox-sensitive priority pollutants. Bioaugmentation of the barrier matrix with autotrophic denitrifiers can shorten the acclimation period required for indigenous bacteria to colonize the Fe(0) surface. Alternatively, an indigenous hydrogenotrophic consortium is likely to eventually develop around an H_2-producing Fe(0) barrier. Once microbial populations are established, the bacteria may enhance the treatment process by reducing the amount of nitrite acumulating as an intermediate of abiotic nitrate reduction, and improve Fe(0) corrosion kinetics by removing the passivating H_2 layer from the Fe(0) surface. Thus, appropriate bacteria can enhance both the rate and extent of transformation and yield a more favorable end product distribution.

REFERENCES

Challis, B. (1973). "Rapid Nitrosation of Phenols and its Implications for Health Hazards from Dietary Nitrites." *Nature.* 244: 466.

Daniels, L., N. Belay, B. Rajagopal, and P. Weimer. (1987). "Bacterial Methanogenesis and Growth from CO_2 with Elemental Iron as the Sole Source of Electrons." *Science.* 23: 509-511.

Huang, Chin-Pao, Hung-Wen Wang, and Pei-Chun Chiu (1998). "Nitrate Reduction by Metallic Iron." *Water Research.* 32: 2257-2264.

Kapoor, A. and T. Viraraghavan. (1997). "Nitrate Removal from Drinking Water-Review." *Journal of Environmental Engineering.* 123(4): 371-380.

Powell, R. M and R. W. Puls (1997). "Proton Dissolution of Intrinsic or Augmented Aluminosilicate Minerals for in Situ Contaminant Remediation by Zero-Valence-State Iron." *Environmental Science and Technology.* 31: 2244-2251.

Till, B. A., L. J. Weathers, and P.J. Alvarez, "Fe(0)-Supported Autotrophic Denitrification." *Environ. Sci. Technol.* (1998), 32, 654-659.

Weathers, L. J., G. F. Parkin, and P. J. J. Alvarez. (1997). "Utilization of Cathodic Hydrogen as Electron Donor for Chloroform Cometabolism by a Mixed Methanogenic Culture." *Environmental Science and Technology.* 31(3): 880-885.

Yoshinari, T. and R. Knowles. (1976). "Acetylene Inhibition of Nitrous Oxide Reduction by Denitrifying Bacteria." *Biochemical and Biophysical Research Communications.* 69: 705-710.

DENITRIFICATION OF SEPTIC TANK EFFLUENT USING SULFUR/LIMESTONE PROCESSES

Jun Shan and ***Tian C. Zhang*** (University of Nebraska-Lincoln, Omaha, NE)

ABSTRACT: In this study, the feasibility of coupling a conventional lateral field with a sulfur/limestone (S/L) layer to treat nitrate in septic tank effluent was investigated using column reactors to simulate the septic tank soil adsorption system. The effects of different hydraulic loading rates, nitrogen loading rates, the depth of S/L layers and the ratio of S/L to gravel on reactors' performance were investigated. The profiles of ammonium, nitrite, nitrate, sulfate, calcium and other parameters along the depth of the reactors were measured. Significant nitrification was observed in the sand layer. Significant denitrification and moderate production of SO_4^{2-} or Ca_2^{2+} were observed in the sulfur/limestone layer. The result showed that S/L autotrophic denitrification was very efficient for nitrate treatment in septic tank effluent.

INTRODUCTION

Septic tank systems are the second largest source of groundwater nitrate contamination in the U.S. (Canter and Knox, 1988). However, there is no readily available way to solve this problem. The objective of this project is to study the feasibility of using a sulfur limestone autotrophic denitrification (SLAD) system for in-situ treatment of nitrate from septic tank system effluent. In this process, a sulfur/limestone (S/L) layer is introduced below the sand layer of a conventional lateral field. Autotrophic denitrificans (e.g., *Thiobacillus denitrificans*) can use nitrate as the electron acceptor, and elemental sulfur as the electron donor. Limestone is introduced to provide the carbon source and alkalinity for maintaining a neutral pH environment. Such a system is called a Sulfur/Limestone Autotrophic Denitrification (SLAD) System.

MATERIALS AND METHODS

Reactors. Four packed-bed columns, 8 inches (20 cm) in diameter and 4 ft (1.2 m) in height, were built to simulate the lateral field (Figure 1). Each column was filled with a sand layer and a S/L layer. Different heights of S/L layers and sand layers were designed to study the optimal depth of each layer. The S/L layer in columns #1 to #5 had a S/L ratio of 3:1 (v/v) (Table 1). The raw wastewater was the primary effluent from a local wastewater treatment plant. Seeding was achieved by continuously pumping raw wastewater through the SLAD system.

Experimental Design. The first phase was from day 1 to day for feasibility studies. After 150 days, 10 mg/L NH_4^+-N was added into the influent for 10 days, followed by an addition of 20 mg/L NH_4^+-N for another 10 days. This was designed to study the effect of the loading rates on reactor performance.

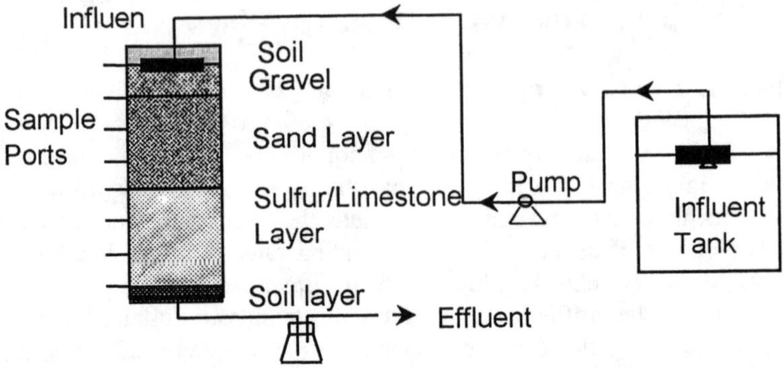

FIGURE 1. Layout of the lab scale septic tank lab system.

After 180 days of operation, column #3 was changed to column #5 by refilling 1 foot (30 cm) of a mixture of S/L (50% v/v) and gravel (50% v/v) and 2 feet (60 cm) of sand. The S/L and gravel mixture was designed to reduce the system capital cost. Column #4 was changed to column #6, a control column filled with gravel and sand only (Table 1).

TABLE 1. Differences between columns.

Column	Sand Layer (ft)	S/L Layer (ft)	Surface Loading Rate ($L/m^2/d$)	Composition of the SLAD Layer
#1	1.5	1.5	32	S/L (v:v = 3:1) only
#2	2	1	160	S/L (v:v = 3:1) only
#3	1.5	1.5	160	S/L (v:v = 3:1) only
#4	1	2	160	S/L (v:v = 3:1) only
#5	2	1	160	50% S/L (v:v=3:1)*
#6	3	0	160	No S/L, sand only

* Another 50% was gravel.

Analytical Methods. Samples were taken every half foot along the column depth to analyze ion concentrations. A Dionex 500 IC/HPLC was used to analyze NO_2^-, NO_3^-, NH_4^+, SO_4^{2-}, and Ca^{2+} concentrations of all the samples. Alkalinity and S^{2-} were analyzed based on the *Standard Method.* COD was analyzed by the HACH method. ORP and pH were measured using the ORP/pH meter. Dissolved Oxygen (DO) was measured using a DO meter. Each data point reported in the paper was the average of five measurements.

RESULTS AND DISCUSSION

Nitrification and Denitrification. The influent total nitrogen was 30 - 50 mg/L. The nitrogen removal was greater than 90% after 1 month of operation (Figure 2). The effluent NO_2^- remained lower than the detection limit (0.1 mg N/L); the

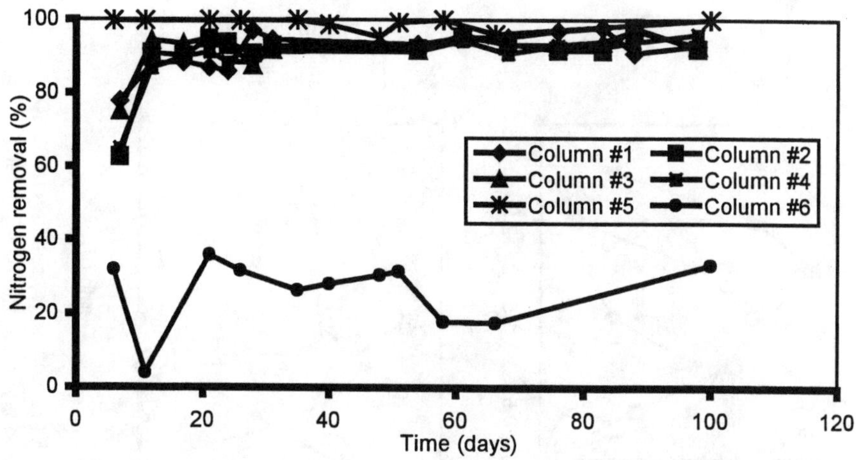

FIGURE 2. Time courses of nitrogen removal. Reactors #5 and #6 were run 6 months after other columns. Introducing the data of these two reactors in this figure is for the purpose of comparison.

effluent NO_3^- was lower than 1 mg N/L; and the effluent NH_4^+ was lower than 4 mg N/L. The control column (#6) had effluent nitrate concentrations of 15 - 28 mg N/L, which was about 4 - 40% nitrogen removal, with an average of 24%.

As wastewater flowed down the sand layers of the columns, NH_4^+ concentrations decreased, while NO_3^- concentrations increased, suggesting that nitrification occurred in the gravel/sand layer (Fig. 3(a) and 3(c)). 65 – 90 % nitrification happened within 1 foot (30 cm) of the gravel/sand layer and 80% happened in the first 1.5 feet (45 cm).

Significant denitrification occurred in the S/L layer (Fig. 3(a)), while the nitrate concentration in the control column remained high, showing that no denitrification happened after nitrification (Fig. 3(f)). Up to 50 mg N/L nitrate removal occurred in the first 0.5 feet (15 cm) of the SLAD layer.

NH_4^+ concentrations in both the SLAD columns and the control column decreased along the depth. However, NH_4^+ concentrations remained at about 2 – 3 mg N/L with a slight increase in the S/L layers (Fig. 3(a)). In the control column, however, the effluent NH_4^+ concentrations could achieve 0 mg/L N (Fig. 3(f)). This suggests that the increase of NH_4^+ concentrations in the S/L layer is probably caused by the anaerobic/reductive environment, or by the decay of denitrificans in the S/L layer.

Column #5, which had 50% gravel in the S/L layer, also achieved a low effluent NO_3^- concentration (data not shown). Other experiments conducted in the authors' research group showed that mixing S/L with up to 67% of gravel could achieve denitrification well (Flere, 1997). Thus the sulfur limestone dosage can be greatly reduced to further lower the cost.

Sulfate Production. In all the S/L columns, sulfate production was observed in the first half feet of the S/L layer, where nitrate removal was also significant.

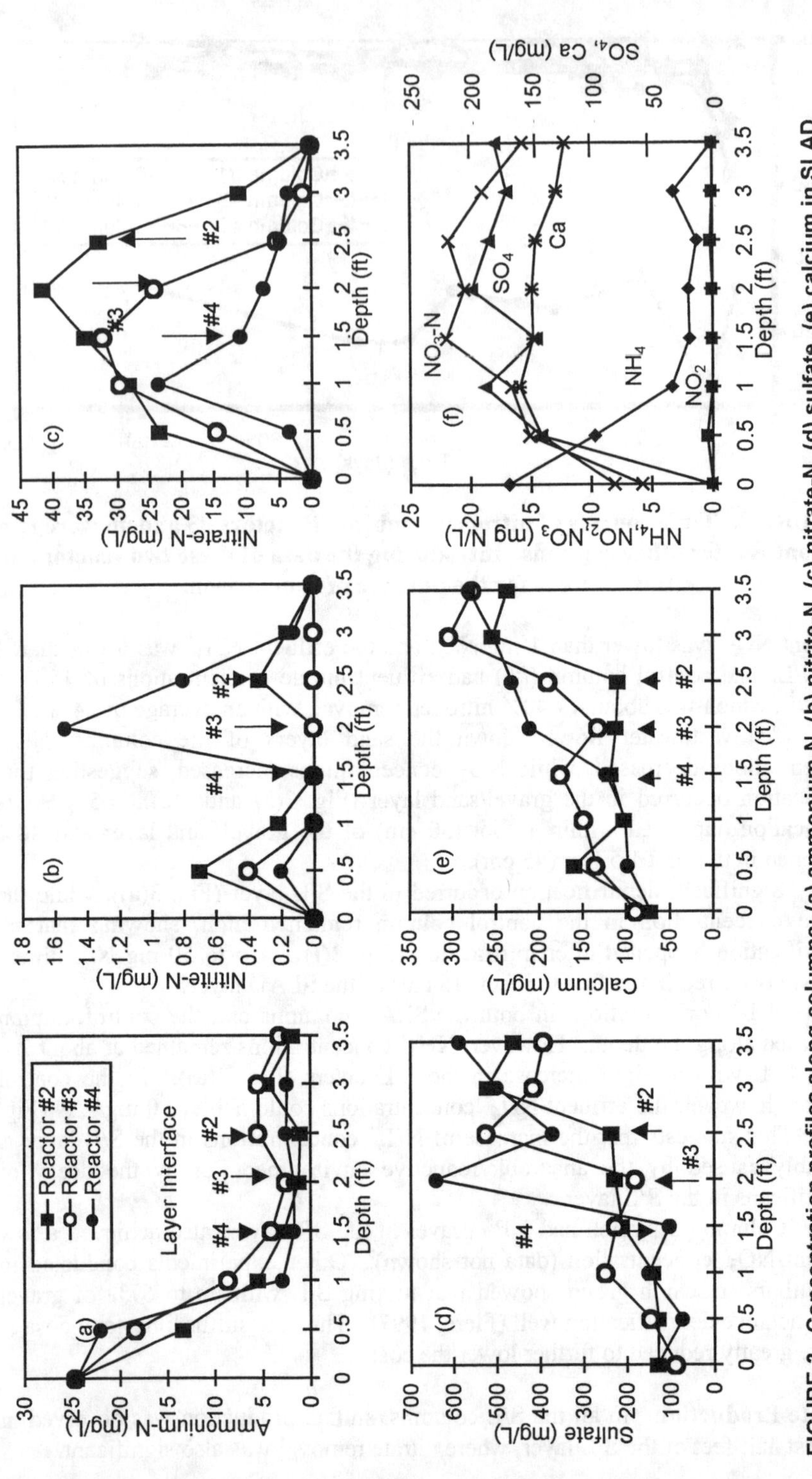

FIGURE 3. Concentration profiles along columns: (a) ammonium-N, (b) nitrite-N, (c) nitrate-N, (d) sulfate, (e) calcium in SLAD columns, and (f) ion profiles in the control column. Arrows indicate the interfaces of the sand layer and the SLAD layer.

Effluent sulfate concentrations ranged from 300 to 400 mg/L. There was no significant difference in sulfate production between different columns.

The stoichiometric equation indicates that 7.5 mg SO_4^{2-} will be produced for every 1 mg nitrate-N reduced (Batchelor and Lawrence, 1978a, 1978b). In this study, it was found that the ratio of sulfate produced to NO_3^--N removed ranged from 4.0 to 9.3, with an average of 7.53.

pH, Alkalinity and Calcium Production. pH profiles along the columns were stable, ranging from 6.4 to 7.6. This obviously is due to the presence of limestone. The alkalinity in influent and effluent were 300 - 400 mg $CaCO_3$/L in all the S/L columns.

An increase in calcium was observed in all the S/L columns, especially in the first half feet of the S/L layer, where sulfate production was also significant. The effluent hardness ranged from 400 to 600 mg $CaCO_3$/L. The theoretical value of the ratio of hardness production to nitrogen removed is 9.14. The average ratio (8.93) obtained in this study is very close to this value, indicating that the calcium produced (hardness increase) is mainly due to limestone dissolution. The hydrogen ions generated from the autotrophic denitrification process are neutralized by the OH^- ions produced in reactions related with limestone dissolution.

There was no difference in sulfate and calcium production between columns #2 and #5. Column #2 had a 100% S/L layer, while column #5 had a layer of 50% gravel plus 50% S/L. Both had a 1 foot S/L layer. This is reasonable because the nitrogen loading rates in both columns were the same, and the removal efficiencies were the same.

DO and ORP. The DO concentrations immediately below the top soil (0.5 feet) were about 3 – 4 mg/L due to the penetration of oxygen from the top soil. The DO concentrations decreased with an increase in the reactor depth, indicating DO consumption by COD removal and nitrification. The DO concentrations were < 1 mg/L in the SLAD layers. No nitrification can take place at such a low DO environment. However, SLAD layers did not necessarily lower the DO because DO in the lower part of the control column was also very low.

ORP profiles shows the difference between the control column and the S/L columns. In all of the gravel/sand layers, the ORP increased gradually, from –110 mV of the raw water to about 150 mV. Once in the S/L layer, the ORP dropped to about 0 to –50 mV, showing a reductive environment, while the ORP of the control column still increased to about 200 mV.

Loading Studies. When the gravel/sand layer could not oxidize all of the ammonium, the NH_4^+ concentration did not decrease in the S/L layer. This suggests that no nitrification took place in the S/L layer, presumably due to the anaerobic environment in the S/L layer. When the total nitrogen in the raw wastewater was approximately 70 - 80 mg TN/L, the effluent NH_4^+ reached 2.0, 7.5, and 10.5 mg N/L in columns #2, #3, and #4, respectively. This suggests that column #4 with a 1 foot (30 cm) sand layer cannot guarantee the complete nitrification of 80 mg/L ammonium, while column #2 can still perform very well due to the longer sand layer. Since a typical concentration of TN in septic tank effluent is 15 - 80 mg/L, at least two feet (60 cm) of the sand layer is required in the soil adsorption system.

Fed with different initial NH_4^+ concentrations, column #2, which had a one foot (30 cm) S/L layer, reached almost complete denitrification. After reaching steady state, the highest NO_3^- concentration in the column #2 effluent remained much lower than 1 mg/L. These results suggest that a 1 foot (30 cm) S/L layer is enough for an initial TN concentration as high as 80 mg TN/L, while the sand layer must have a length of 2 feet (60 cm) to complete nitrification.

Considering that the standard depth of the lateral field is 2 - 4 feet (60 – 120 cm, Canter and Knox, 1988), the introduction of a S/L layer underneath the sand layer will not significantly increase the required depth of the septic tank system, and thus, will not introduce extra site selection criteria and capital cost.

The effluent sulfate and calcium increased with an increase in the influent NH_4^+ concentration. The production of sulfate and calcium also followed the aforementioned stoichiometric equations.

SUMMARY

(1) The SLAD system, combined with the soil absorption system, can remove nitrate from septic tank effluent effectively. (2) Nitrification occurs mainly in the sand layer, while denitrification takes place in the S/L layer. (3) Sulfate and calcium production was observed and followed the stoichiometric equations.

REFERENCES

Batchelor, B., and A.W. Lawrence. 1978a. "Autotrophic denitrification using elemental sulfur." *J. WPCF.* 50(8):1986-2001.

Batchelor, B., and A.W. Lawrence. 1978b. "A kinetic model for autotrophic denitrification using elemental sulfur." *Water Research.* 12:1075-1084.

Canter, L. W., and R. C. Knox. 1988. *Septic Tank System Effects On Groundwater Quality.* Lewis Publishers, Inc. MI.

Flere, J.M. 1997. "Remediation of Nitrate-Contaminated Water Using Sulfur and Limestone Autotrophic Denitrification Processes." M.S. Thesis, University of Nebraska-Lincoln, 1997.

Shan, J. 1998. "Sulfur/Limestone Autotrophic Denitrification Processes for Nitrate Treatment in the Septic Tank Effluent." M.S. Thesis, University of Nebraska-Lincoln, 1998.

LABORATORY AND FIELD TESTING OF ENGINEERED DENITRIFICATION

Sheldon N. Nelson (Chevron Research and Technology Company, Richmond, CA)
Kirk T. O'Reilly (Chevron Research and Technology Company, Richmond, CA)
Miriam L. Barbonio (Chevron Chemical Company, San Ramon, CA)
Gerald Vogt (CH2M Hill, Redding, CA)

ABSTRACT: A four-part laboratory and field test was conducted to determine the applicability of an engineered denitrification strategy to address impacted soil and groundwater beneath a former fertilizer storage facility in the Central Valley of California. The study involved batch testing of nitrate impacted soil, column tests to simulate the treatment of dissolved nitrate in the saturated zone, and a field application to address residual nitrate remediation in the vadose zone. In summary, the study results indicate that an engineered reductive denitrification strategy is effective for addressing nitrate impacted soil and groundwater if carbon is supplied and specific conditions are maintained.

INTRODUCTION

The soil and groundwater beneath a former storage facility in the Central Valley of California have accumulated elevated levels of ammonium and nitrate due to the packaging and storage of ammonium nitrate fertilizer. Biological denitrification has been proposed as a potential in situ remedial option at the property. Laboratory and field trials were conducted to determine the efficacy of using readily available low cost carbon sources to stimulate denitrification in site soil and groundwater.

Many soil bacteria can substitute nitrate for oxygen in the oxidation of organic compounds. This process, called denitrification, leads to the conversion of nitrate to nitrogen. The conditions required for denitrification are the availability of degradable organic compounds, the presence of nitrate, the appropriate microbial community, and a lack of available oxygen. A wide variety of carbon sources are suitable for stimulating this process. If sugar is used, denitrification can be characterized by the following equation:

$$\underset{\text{(sugar)}}{C_6H_{12}O_6} + \underset{\text{(nitrate)}}{4\,NO_3^-} \rightarrow 6\,CO_2 + 2\,N_2 + 6\,H_2O$$

Oxygen depletion occurs quickly in water saturated or biologically active subsurface soils, thereby inducing localized anaerobic conditions. It is possible to stimulate denitrification by flooding soils to reduce oxygen transfer into the soil from the air. The denitrification process can be further stimulated by adding a readily degradable carbon (food) source for the denitrifying bacteria. The rapid oxidation of the supplied carbon source by aerobic bacteria will reduce any oxygen present in the subsurface soil sufficiently for denitrifying activity to commence.

METHOD

Four 5-gallon buckets of soil were collected from the Central Valley test site. Initial nitrate measurements indicated that three of the buckets had similar nitrate concentrations (approximately 300 ppm $N\text{-}NO_3$), while the fourth had less than 50 ppm. The contents of the three similar buckets were screened through a ¼" mesh and mixed together using a small cement mixer. The mixed soil was used for the series of laboratory tests described below.

PHASE I: BATCH TEST

Objectives

- Evaluate denitrification as a treatment option for the removal of nitrate from soil.
- Determine if two available carbon sources, dilute sugar water and raw molasses solution, will promote denitrification in the sample test soil that contains nitrate at concentrations of 300 and 1000 ppm $N\text{-}NO_3$.
- Calculate the denitrification rate.

Procedure

Two carbon sources, dilute sugar water (DSW) and molasses were used in the batch tests. Soil that was not amended with carbon was used as a control. Both carbon sources were obtained from an animal feed plant situated near the site. The soil samples used in the test contained either nitrate concentrations of 300 ppm $N\text{-}NO_3$ (as collected from the site), or were spiked to a concentration of 1000 ppm $N\text{-}NO_3$. Sufficient carbon source was added to give a 3:1 COD ratio to residual nitrate. After the soil was mixed with the appropriate carbon source and nitrate, 1 kg samples were placed in 1 L jars. To mimic the anaerobic conditions expected in the COD rich soil samples, all but one of the tests were run in sealed jars with a N_2 headspace. One jar was left open to investigate the effect of drying and O_2 exposure.

Results – Phase I

- The results indicate that the appropriate species of microorganism were present in the tested soil to actively denitrify, in the presence of sufficient carbon source.
- After a short time lag, which allowed the population of microorganisms to grow to a sufficient level, nitrate concentrations were observed to decline in the batch test samples. Utilization rates as high as 62 ppm $N\text{-}NO_3$ per day were measured.

- Five to 25 percent of the initial nitrate remained in the soil at the end of the test. This could have been due to the depletion of 80-90% of the added COD, or to a lack of mixing within the system.
- The lack of nitrate reduction in the jar exposed to the air suggests that denitrification will not occur in well drained, aerated soils.
- The lack of nitrate treatment in the control jars with no additional COD indicates that the naturally occurring carbon source is generally not suitable for the resident bacteria to induce denitrification.

PHASE II: COLUMN STUDY

Objectives

The injection of carbon rich water was proposed as a method for stimulating denitrification in groundwater. This option was tested in saturated soil columns.

Procedure

Soil columns were constructed from PVC plastic piping. The columns were 4 inches in diameter and 24 inches in length. Sample ports were placed at 5 locations in the sidewall of each column (2.5", 5", 10", 15" from the surface, and column out). Screened wide bore syringe needles inserted to the mid-point of the column at each sample point was used to extract water from the soil.

Each column contained 20" of soil. The top 10" of the column was packed with site soil with a nitrate concentration of 300 mg/Kg N-nitrate. The site soil mix was treated with lime (5 g/Kg) to raise the pH, and super triple phosphate (0.2 g/Kg) as a nutrient source. The bottom 10" of the test column was a nitrate free, low COD, sandy soil. After packing, the soil was saturated from the bottom with water to minimize leaching while inducing anaerobic conditions. Approximately 2.8 kg of site soil was used in each column.

To mimic ground water flow, 75 ml of water was added to the top and removed from the bottom of the columns each day. The hydraulic retention time within the soil column was about 2 weeks. After 11 days, dilute sugar water or ethanol (COD 2400 ppm) was added to the feed of two of the columns . Water added to a control column did not contain any dissolved COD. After an additional 2 weeks, nitrate (300 ppm N-NO_3) was added to the feed of all the columns.

Results

- When nitrate and COD rich water were added to the columns, the concentration of nitrate dropped from approximately 300 ppm to near detection limits within the top 2.5 inches of soil.
- A calculated nitrate utilization rate exceeding 150 ppm/day was determined, based on the observed vertical water velocity within the column of less than 2 days between nitrate addition and the uppermost sample port.

- Little change in nitrate concentration was noted in the control column, suggesting that a readily bioavailable carbon source is essential to induce denitrification.

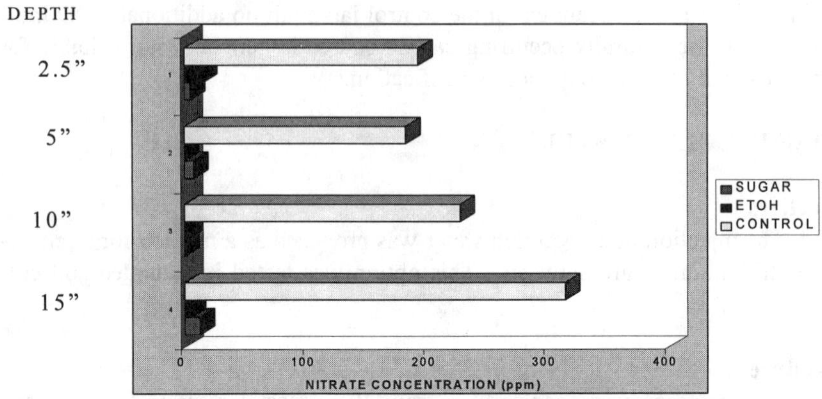

FIGURE 1. Nitrate concentration with depth after 1 month of nitrate addition

PHASE III: PASSIVE BARRIER FOR DISSOLVED NITRATE TREATMENT

Objective

A second potential method for groundwater treatment is the construction of a passive barrier that contains a solid carbon source. The strategy would be to add sufficient carbon to support denitrification for a number of years. This was tested in a saturated soil column.

Procedure

A variety of low cost animal feeds were evaluated as potential carbon source material. Canola meal pellets were selected for the column test since they released the lowest concentration of dissolved nitrate of the products tested.

A column used in the previous test was repacked for this study. The top and bottoms of the column contained clean soil. To mimic a passive biobarrier trench, the central 10 inches were packed with a sand and gravel mix containing a 20% (by volume) canola meal pellets. Water containing 500 ppm $N\text{-}NO_3$ was added to the column at a rate 75 ml per day. The nitrate concentration at each sample port was analyzed weekly. The test was run for over 200 days.

Results

- Denitrification started within three weeks.
- Most of the denitrification occurred between the 1st and 2nd sample port.

- Incomplete nitrate reduction characterized the test, possibly due to a lowering of pH to conditions which were too acidic for maximum denitrification.
- When the influent nitrate concentration was lowered to 200 ppm, the effluent concentration decreased to about 50 ppm.

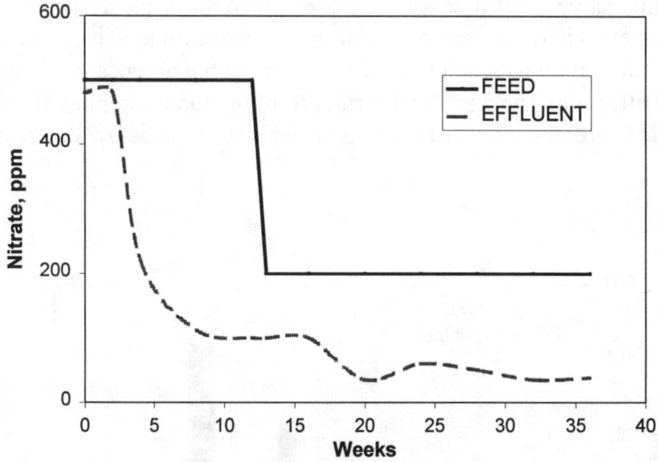

FIGURE 2. Feed and effluent concentration of the passive barrier column test.

PHASE IV: FIELD TRIAL – DENITRIFICATION OF NITRATE-IMPACTED SOIL – CENTRAL VALLEY SITE

Objective

To field test the ability to stimulate in situ denitrification in shallow soil using a mixture of dissolved (sugar) and solid (canola meal pellets) carbon sources.

Procedure

This test was conducted in the winter of 1997-98. Soil at the site was screened to identify an area of elevated nitrate concentration. The upper two foot section of a 20 by 8 foot test plot was amended with 2200 pounds of canola meal pellets, 300 pounds of sugar, 150 pounds of oyster shell agricultural lime, and 5 pounds of triple super phosphate fertilizer. Soil was removed from the plot, mixed with the amendments and reintroduced into the excavation. The soil of the test plot was kept saturated during most of the duration of the experiment.

Soil borings were collected at three depths to determine the impact of carbon addition within (1 foot depth), and below (3 and 6 foot depth), the amended area. Samples were collected at 0, 3 and 5 months. The soil was analyzed for nitrate and COD.

Results

Five of the six soil analytical data pairs collected at the one foot depth exhibited a marked decline in the measured concentration of residual nitrate (Figure 3), suggesting active denitrification processes. In contrast, the deeper 3 foot and 6 foot samples exhibited no clear reduction in residual nitrate values in a comparison of 0, 3 and 5 month samples; indicating a probable absence of significant induced denitrification. The analytical results from the three sampling depths imply that the added carbon source does not appreciably migrate below the depth of initial mixing.

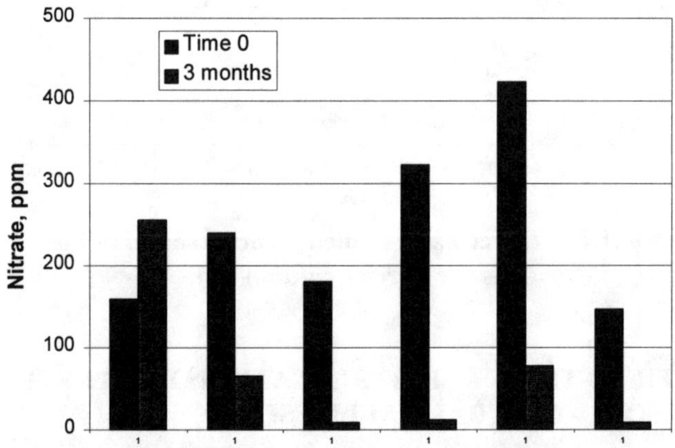

FIGURE 3. Soil nitrate concentrations at 6 locations in the field plot at Time 0 and 3 months.

CONCLUSIONS:

The series of tests described above indicates the following conclusions and potential applications for the remediation of fertilizer impacted sites:

- Engineered, biological denitrification of soil and groundwater may be induced by the addition of a readily-available carbon source, to an anaerobic environment, under neutral to slightly alkaline pH conditions.
- As noted under field conditions, the effect of the carbon source is only detected within the actual amended column of soil. Dissolved carbon source does not apparently migrate vertically within the soil column in sufficient quantity to encourage denitrification.

REMOVAL OF NITRATE FROM AQUEOUS STREAMS

Ernst A. Stadlbauer, Markus Röhricht, Harald Happel, Bernd Weber, Jochen Eberheim, Uwe Sauerland and Sebastian Bojanowski (University of Applied Sciences, Giessen, GERMANY)
Joachim Esch (Schunk Company, Heuchelheim, GERMANY)

ABSTRACT: A novel approach for nitrate removal from aqueous streams combines reverse osmosis with electro-catalytical nitrate reduction. It is shown that nanofiltration membrane elements (loose reverse osmosis elements) can retain nitrate to more than 90% under certain conditions due to the Donnan effect. The concentrate of the reverse osmosis process, which accounts for about 25% of the raw water, can be reduced to dinitrogen by an electro-catalytical method. In situ produced hydrogen (H) reacts at a catalytically active high area cathode with nitrate to form dinitrogen (N_2). Depending on the catalytic system used specific removal rates for nitrate (NO_3^-) and nitrite (NO_2^-) are up to 145 mg/m^2×min.

INTRODUCTION

Widespread use of nitrogen-based fertilizers in agriculture is mainly responsible for nitrate pollution in ground water. Due to hazards to human health (especially for babies), there are limits for maximum allowable nitrate concentration in drinking water. WHO regulations show c(N-NO_3^-) = 10 mg/L corresponding to 44 mg/L nitrate. European community specifies a limiting value of c(NO_3^-) = 50 mg/L. Presently, in many ground water reservoirs the 50 mg/L limit is exceeded. Consequently this water cannot be used as drinking water without appropriate treatment. Both physical techniques such as ion exchange, reverse osmosis and electrodialysis as well as biological methods are applied. This purification increases municipal water charges by 0.75 to 1.0 \$/m^3. Reduction to nitrogen gas is the foremost advantage of in situ or on-site denitrification (Eichhorn D., 1991). A variety of techniques including immobilized enzymes are available (B. Herrling et. al., 1994). As far as treatment of drinking water is concerned, surveillance for potential bacterial contamination is necessary.

A drawback of membrane separation processes is that nitrate is not degraded. Instead, it is given back to the environment in higher concentration. In average reverse osmosis or nanofiltration applications, the water recovery is approx. 75 % meaning that 25% of the raw water is turned into waste water. To avoid additional disposal costs, on-site treatment of this concentrate is advisable.

In a novel approach we use nanofiltration to remove nitrate efficiently from raw water sources, while an **e**lectro-**c**atalytic process for **n**itrate reduction (**ECN**) is applied to treat the nanofiltration concentrate (Figure 1). This procedure increases the overall water recovery to nearly 100 percent. High nitrate concentration generated by nanofiltration increases the efficiency of the electro-catalytic degradation of nitrate.

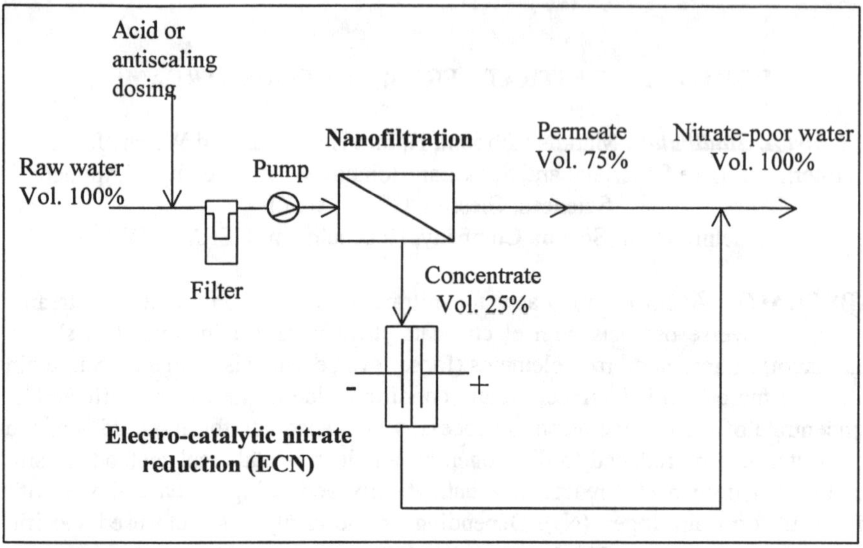

FIGURE 1. Flow sheet of a nanofiltration plant for on-site nitrate removal coupled with an electro-catalytic nitrate reduction system.

The electro-catalytical process uses an electrochemical cell where hydrogen ions are transported from an anolyte compartment through a membrane into the catholyte section. At a high surface area cathode, activated hydrogen (H) is formed. In the presence of a suitable catalyst, nitrate is reduced to dinitrogen described by the stoichiometric equation: $2\ NO_3^- + 10\ H \rightarrow N_2 + 2\ OH^- + 4\ H_2O$.

In order to use the treated water as potable water, further steps such as mixing with other water sources, pH-adjustment or chlorination (disinfecting) may be necessary.

Objective. The objective of this study is to optimize the membrane separation and electro-catalytic nitrate reduction, first by working separately on each process and then combining both to one system. The already well known reverse osmosis technique can be optimized by using nanofiltration membrane elements which have a higher flux then conventional reverse osmosis elements. Commercially available membrane elements were tested on nitrate rejection and water flux. Moreover, the ECN - process demands a selective catalyst with high activity and good stability.

Site Description. The nitrate contaminated well is situated at Reiskirchen, Hesse, approx. 80 km north of Frankfurt, Germany. Contamination is due to extensive farming in the past. Well depth is about 10 m showing a fluctuation in nitrate concentration from 45 to 55 mg/L. Up to now raw water runs through ion exchange modules before distribution into the municipal water supply.

MATERIALS AND METHODS

Nanofiltration experiments. A pilot scale reverse osmosis system suitable for one 2.5" spiral wound membrane element was used for all experiments. Maximum water flow showed 700 L/min. Nine different reverse osmosis (high flux) and nanofiltration membrane elements, each having an active membrane area of 2.5 m², were tested. The water sources used are listed in Table 1.

TABLE 1. Raw water sources for the nanofiltration experiments.

Water source	Nitrate conc.	pH-value	Location of experiment
Softened municipal water	16 mg/L	7.0	Laboratory
Artificial nitrate water	45 mg/L	5.5	Laboratory
Nitrate contaminated well	45-55 mg/L	5.5	On site (Reiskirchen)

Electro-catalytical nitrate reduction experiments. Water to be treated is fed into the catholyte compartment of an electrochemical cell using a cathode separated from an anolyte section by a membrane such as Nafion® 300 (DuPont). The cathode consists of carbon/carbon (C/C, fibrefoil CF 262, Schunk) or graphite felt (GF 1) impregnated with copper and palladium (2:98). Copper/Tin alloys are also suitable (Stadlbauer et al., 1997). Preferably, graphite plates are used as anodes. Water or a non oxidizing acid such as dilute sulfuric acid flows through the anolyte section (Figure 2). Temperature, pH, flow rates and nitrate are measured on-line (Endress & Hauser). Controls for nitrate and nitrite are analyzed by anion chromatography (AS 2 column, Dionex).

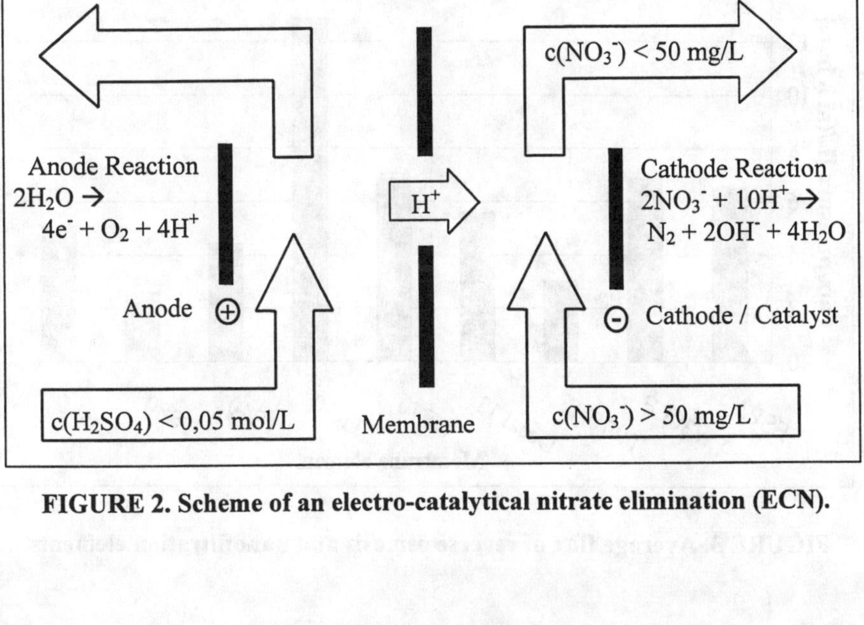

FIGURE 2. Scheme of an electro-catalytical nitrate elimination (ECN).

RESULTS AND DISCUSSION

Nanofiltration experiments. In Figures 3 and 4, flux and nitrate retentions are shown for the nine elements, respectively. The flux varies from 3 to 7 L/m²×h×bar with one exception at 13 L/m²×h×bar. It shows no general difference between those modules which are declared reverse osmosis elements (no. 1-4) and those elements which are declared nanofiltration elements by the manufacturer (no. 5-9). However, looking at the nitrate retention, the four reverse osmosis membrane elements show an efficiency above 90% for all water sources which is generally higher than for the nanofiltration elements. The nitrate retention of nanofiltration elements depends highly on the water source. It can even become negative as in the case of the TR 20 element working with softened municipal water.

Nanofiltration membrane elements are often defined as "loose" reverse osmosis elements showing retentions above 90% for divalent ions such as SO_4^{2-} and poor retentions for monovalent ions such as NO_3^-. Therefore, nanofiltration does not seem appropriate for the retention of a monovalent ion such as nitrate. However, as results show, even nanofiltration membrane elements can reduce nitrate by more than 90% under certain circumstances. This is due to the Donnan effect or Donnan equilibrium which tries to equalize electrical charges on both sides of the membrane. (Without this effect, the permeate could become e.g. positively charged because most monovalent ions such as sodium Na^+ pass through the membrane while the concentrate will become negatively charged because all divalent anions such as sulfite SO_4^{2-} are retained). Depending on water composition, the Donnan equilibrium can have positive or negative effects on nitrate retention as shown for the elements TR 20 and TS 40 in figure 4.

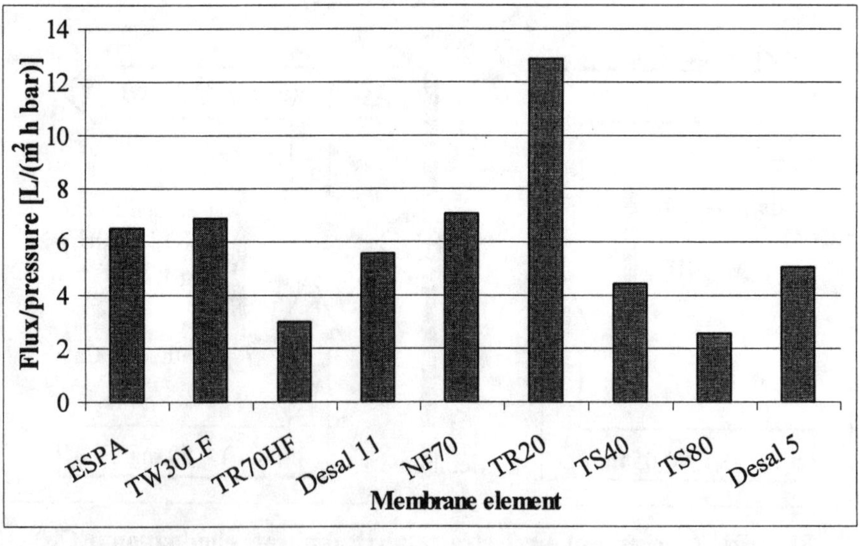

FIGURE 3. Average flux of reverse osmosis and nanofiltration elements.

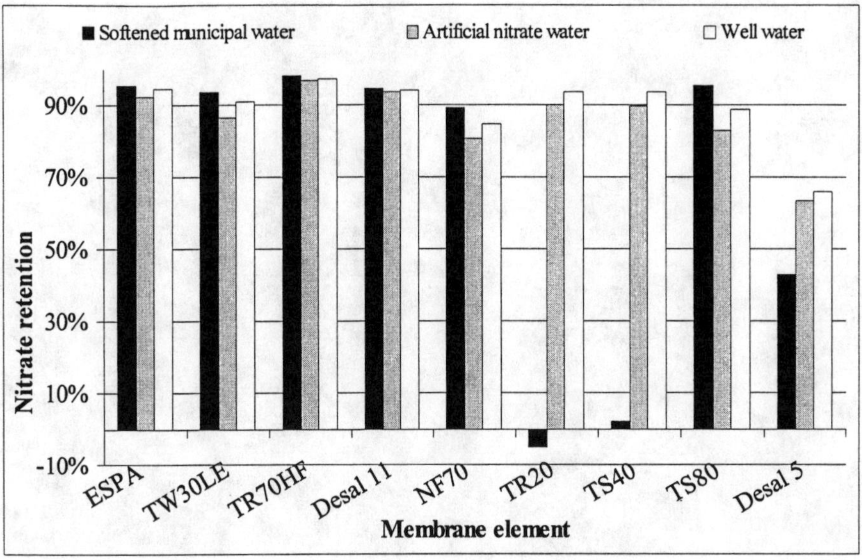

FIGURE 4. Nitrate retention of reverse osmosis and nanofiltration elements using three different water sources.

ECN Process. Table 2 shows results of a screening for supported catalysts able to eliminate both nitrate and nitrite in aqueous solution at room temperature. The most favorable reaction conditions are characterized by pH 3 – 5 at current densities $\rho < 20$ A/m^2. In such a medium, production of ammonium ions (NH_4^+) is minimized.

TABLE 2. Solid catalytic systems investigated for the ECN process. Current densities $\rho = 10 - 20$ A/m^2; current efficiency $\eta = 80\text{-}90\%$; pH 3-5.

Support / Cathode	Catalytically active component	Specific removal rate a(X) mg/m^2×min	Concluding reaction
Cu	Cu	35	$NO_3^- \rightarrow NO_2^-$
Ti	Pd	53	$NO_2^- \rightarrow N_2$
Pt	Cu/Pd (2:98)	145	$NO_3^- \rightarrow N_2$
Steel (Porostar)	Cu/Pd (2:98)	93	$NO_3^- \rightarrow N_2$
Graphite felt	Cu/Pd (2:98)	77	$NO_3^- \rightarrow N_2$
C/C plate	Cu/Pd (2:98)	68	$NO_3^- \rightarrow N_2$
Bronze	Cu/Sn	112	$NO_3^- \rightarrow N_2$

With the exception of bronze, these catalysts were prepared under potential deposition. To enable successful plating, the main problem is to create a close interacting between the two metals or graphite and the metal (Figure 5). Massive Cu/Sn alloys show advantages in this respect.

FIGURE 5. SEM of Pd-deposition on graphite felt.

Evaluation of data indicates that the combination of nanofiltration and electro-catalytic reduction has the potential for in situ remediation of groundwater containing nitrate. Organic contaminants dissolved in the concentrate require further purification. In addition, the electro-catalytic reactor offers an effective tool for removing nitrate from various technical fluids such as regenerants from ion exchange, lubricating coolants or oxygen from boiler feed.

REFERENCES

Eichhorn, D. 1991. *Wasserbehandlung im Grundwasserleiter – Denitrifikation – in situ Test Tännicht*. Abschlußbericht zum BMFT-Vorhaben MFT-7270-30F202200, Institut für Wassertechnologie, Dresden, GERMANY.

Herrling, B., J. Stamm, E. J. Alesi, G. Bott-Breuning, and S. Diekmann. 1994. "In Situ Bioremediation of Groundwater Containing Hydrocarbons, Pesticides, or Nitrate Using Vertical Circulation Flows (UVB/GZB Technique)." In R. E. Hinchee (Ed.), *Air Sparging for Site Remediation*, pp. 56-80. CRC Press, Boca Raton, FL.

Stadlbauer, E.A., U. Sauerland, B. Weber. 1997. *Elektrokatalytische Nitratelimination - EKN-Verfahren*. Abschlußbericht zum Förderprojekt der Deutschen Bundesstiftung Umwelt, Osnabrück, DBU 07874, Fachhochschule Giessen-Friedberg, Labor für Entsorgungstechnik, Giessen, GERMANY.

BIOREMEDIATION IN SITU OF SOILS CONTAMINATED WITH RADIOACTIVE ELEMENTS

S.N. Groudev (University of Mining and Geology, Sofia, Bulgaria)
P.S. Georgiev (University of Mining and Geology, Sofia, Bulgaria)
I.I. Spasova (University of Mining and Geology, Sofia, Bulgaria)
K. Komnitsas (National Technical University of Athens, Greece)

ABSTRACT: An experimental plot containing soil heavily contaminated with radioactive elements (uranium, radium, thorium) and toxic heavy metals (copper, cadmium, lead) was treated by means of an in situ biotechnological method based on the activity of the indigenous soil microflora. The soil was characterized by a positive net neutralization potential and the soil pH was in the slightly alkaline pH range (from 8 to 9). The pollutants were located mainly in the upper soil horizons (mainly in the horizon A) and was present mainly in forms susceptible to biological and/or chemical leaching. The treatment was connected with the dissolution of the pollutants in the upper soil horizons and their transfer into the deeply located soil horizons (mainly to the horizon B_2 where they were immobilized as different insoluble compounds. The dissolution of contaminants was connected with the activity of both heterotrophic and chemolithotrophic aerobic microorganisms and the immobilization were due mainly to the anaerobic sulphate-reducing bacteria. The activity of these microorganisms was enhanced by suitable changes in the levels of some essential environmental factors such as water, oxygen and nutrient contents in the soil.

INTRODUCTION

Some agricultural lands located in the Vromos Bay area, near the Black Sea coast, Southeastern Bulgaria, have been contaminated with radioactive elements (uranium, radium, thorium) and toxic heavy metals (copper, cadmium, lead) as a result of mining and mineral processing of polymetallic ores. Laboratory experiments carried out with soil samples from these lands revealed that an efficient remediation of the soils was achieved by an in situ treatment method based on the activity of the indigenous soil microflora. (Groudev et al., 1999). The treatment was connected with the dissolution of the contaminants in the upper soil horizons and their transfer into the deeply located soil horizons (mainly to the horizon B_2) where they were immobilized as different insoluble compounds. The dissolution of contaminants was connected with the activity of both heterotrophic and chemolithotrophic aerobic microorganisms and the immobilization were due mainly to the anaerobic sulphate-reducing bacteria. The activity of these microorganisms was enhanced by suitable changes in the levels of some essential environmental factors such as water, oxygen and nutrient contents in the soil.

In 1998 the above-mentioned treatment was applied under real field conditions in a heavily contaminated experimental plot of land located in the Vromos Bay area. Data from this study are presented in this paper.

MATERIALS AND METHODS

The experimental plot had a rectangular shape and was 400 m^2 in size (20 m x 20 m). The soil profile was 90 cm high (horizon A 25 cm, horizon B 55 cm, horizon C 20 cm). Leach solutions containing carbonate ions and some dissolved organic compounds were used to facilitate the dissolution of the pollutants. The upper soil layers were ploughed up periodically to enhance the natural aeration. Water solutions of dissolved organic compounds and ammonium phosphate were injected through vertical pipes to the deeply located soil layers (horizon B_2) to enhance the action of the anaerobic sulphate-reducing bacteria.

The flowsheet included also a collection system for the soil drainage solutions. These solutions were treated to remove the residual amounts of toxic elements, non-precipitated in the deeply located soil horizons, by means of a passive system of the type of constructed wetlands.

All analytical procedures were carried out by methods described previously (Groudeva et al., 1993).

RESULTS AND DISCUSSION

Data about the chemical composition and some essential geotechnical parameters of the soil are shown in Table 1. The concentrations of contaminants were higher in the upper soil layers (mainly in the horizon A) (Table 2). The contaminants were present mainly in forms susceptible to biological and/or chemical leaching but considerable portions were present as the relevant inert fractions, which are refractory to solubilization. The soil was characterized by a positive net neutralization potential and the soil pH was in the slightly alkaline pH range (from 8 to 9).

The permeability of the soil was high and rainwater infiltrated into and created conditions favorable for the dissolution of elements. The pore water quality was poor, with high concentrations of contaminants such as radioactive elements (uranium, radium, and thorium) and some toxic heavy metals (copper, cadmium, lead). Prior laboratory experiments with soil samples from the experimental plot treated in this study had shown that the dissolution of contaminants was connected with the activity of the indigenous soil microflora. The analysis of this microflora revealed that it included a rich variety of microorganisms (Table 3). In the upper soil layers the aerobic heterotrophic bacteria were the prevalent microorganisms. Their total number was higher than 10^8 cells/g dry soil. Bacteria related to the genera Pseudomonas and Bacillus were the most numerous in these microbial communities. The fungi were low in number. Some chemolithotrophic bacteria able to oxidize S^0 and soluble inorganic sulphur compounds at neutral and alkaline pH were also present. Thiobacillus thioparus was the main species among these chemolithotrophs but Thiobacillus neapolitanus, the anaerobic Thiobacillus denitrificans and some mixotrophic bacteria (mainly such related to the species Thiobacillus novellus) were also present.

The activity of this microflora was enhanced by suitable changes in the levels of some essential environmental factors such as water, oxygen and nutrient contents in the soil. This was achieved by regular ploughing and irrigation of the soil.

Table 1. Characteristics of the soil used in this study

Parameters	Horizon A (0 - 25 cm)	Horizon B (26 – 70 cm)
Chemical composition (in %)		
- SiO_2	60.6	61.1
- Al_2O_3	14.3	14.0
- Fe_2O_3	12.5	5.90
- CaO-	5.72	2.35
- MgO	3.21	1.31
- K_2O	3.92	3.81
- Na_2O	0.88	1.94
- S total	0.28	0.21
- S sulphidic	0.14	0.10
- humus	3.7	1.4
Bulk density, g/cm^3	1.51	1.63
Specific density, g/cm^3	3.14	3.43
Porosity, %	44	41
Moisture capacity, %	42	39
Permeability, cm/s	5×10^{-2}	5×10^{-2}
pH (H_2O)	7.54	7.65
Net neutralization potential, kg $CaCO_3$/t	53	+ 215

The optimum soil humidity was about 45-50% from the moisture capacity of the soil but periodic flushing with water was needed to remove the dissolved contaminants. Zeolite saturated with ammonium phosphate was added to the soil (in amounts in the range of 2-5 kg/ton dry soil) to provide the microorganisms with ammonium and phosphate ions and to improve the physico-mechanical properties of the soil.

The treatment of the soil was started in March 1998 and at the end of October 1998 it was found that considerable portions of the contaminants were removed from the upper soil horizons and their residual concentrations, with the exception of that of the lead, were lowered below the relevant permissible levels. It was found that the leaching rates markedly depended on the presence of organic compounds and carbonate ions in the leach solutions.

The temperature of the soil also was an essential factor affecting the microbial growth and activity. The highest rates of contaminant solubilization were observed during the summer months (June – August 1998) when the temperature inside the soil profiles was in the range of 19-35^{0}C. However, the process was efficient even at temperatures as low as 10^{0}C (in March 1998) when the well exposed and the most easily soluble forms of the above – mentioned toxic elements were solubilized. The uranium leaching was connected with the microbial production of peroxide compounds, which turned the tetravalent uranium to the hexavalent state. The U^{6+} was then solubilized as uranyl carbonate

Table 2. Toxic elements in the horizon A of the soil before and after the bioremediation

Parameters	U	Ra	Cu	Cd	Pb
Content of toxic elements, ppm:					
- before treatment	35	15	648	7	275
- after treatment	10	6	293	3	132
Bioavailable fraction, ppm:					
a. by DTPA leaching:					
- before treatment	2.60	2.35	268	0.8	0.9
- after treatment	0.14	0.08	53	0.1	0.2
b. by EDTA leaching:					
- before treatment	0.10	0.12	97	< 0.04	1.4
- after treatment	0.01	0.01	17	< 0.004	0.17
Easily leachable fractions – exchangeable + carbonate, ppm:					
- before treatment	2.7	1.9	28	0.2	0.7
- after treatment	0.6	0.5	2.3	<0.04	0.1
Inert fraction, ppm:					
- before treatment	12	6.2	125	4.1	161
- after treatment	8	4.6	105	3.2	105
Toxic elements solubilized during the toxicity test, ppm:					
- before treatment	0.35	0.14	1.92	0.01	0.08
- after treatment	0.07	0.05	0.32	<0.004	<0.004

Note: The contents of radium are given in μg/t and for the radium solubilized during the toxicity test in - ηg/l.

or as complexes with some organic compounds. Radium and the toxic heavy metals were solubilized mainly by means of microbially secreted organic acids.

Considerable portions of the contaminants removed from the upper soil horizons were transferred into the deeply located soil horizons (mainly to the horizon B_2) where they were immobilized as a result of the activity of the anaerobic sulphate-reducing bacteria inhabiting this soil horizon. This activity was enhanced by the water solutions of soluble organic compounds and ammonium phosphate injected to this horizon B_2. Uranium was precipitated as uraninite (UO_2) and most of the radium was adsorbed on the clay minerals present in the horizon B_2. The toxic heavy metals were precipitated mainly as the relevant insoluble sulphides.

The concentrations of pollutants in the soil effluents were low, and in most cases, below the relevant permissible concentrations for waters used in agriculture

Table 3. Concentration of various physiological groups of microorganisms in the soil during the treatment

Microorganisms	Horizon A (0-25 cm)	Horizon B (26-70 cm)
	Cells/g dry soil	
Aerobic heterotrophic bacteria	10^6 - 10^8	10^4 - 10^6
Oligocarbophiles	$10^3 – 10^7$	$10^2 – 10^5$
Cellulose-degrading microorganisms	$10^3 – 10^6$	$10^2 – 10^4$
Nitrogen-fixing bacteria	$10^3 – 10^5$	$10^2 – 10^4$
Nitrifying bacteria	$10^2 – 10^5$	$10^1 – 10^3$
Chemolithotrophic sulphur-oxidising bacteria	$10^3 – 10^7$	$10^2 – 10^5$
Chemolithotrophic iron-oxidising bacteria	$1 – 10^1$	$0 – 10^1$
Anaerobic heterotrophic bacteria	$10^3 – 10^6$	$10^3– 10^7$
Denitrifying bacteria	$10^3 – 10^5$	$10^3 – 10^5$
Anaerobic bacteria fermenting carbohydrates with gas production	$10^3 – 10^5$	$10^3 – 10^6$
Sulphate- reducing bacteria	$10^3 – 10^5$	$10^4 – 10^7$
Fe^{3+} - reducing bacteria	$10^2 – 10^4$	$10^2 – 10^5$
Mn^{4+} - reducing bacteria	$10^1 – 10^3$	$10^2 – 10^4$
Methanogenic bacteria	$0 – 10^2$	$1 – 10^5$
Streptomycetes	$10^3 – 10^6$	$10^2 – 10^4$
Fungi	$10^3 – 10^6$	$10^2 – 10^4$
Total cell numbers	$1 \times 10^8 – 6 \times 10^8$	$1 \times 10^6 – 5 \times 10^7$

and/or industry. These effluents were efficiently treated by the passive system located near the experimental plot.

ACKNOWLEDGEMENTS

The authors would like to greatly acknowledgement the financial support of the European Commission under the Copernicus project entitled “Marine pollution in the Black Sea due to mining activities: risk assessment, development of preventive and remedial actions”, Contract N^0: ERB-IC15-CT96-0114.

REFERENCES

Groudev, S.N., P.S. Georgiev, I.I. Spasova and K. Komnitsas. 1999. “Bioremediation of a soil contaminated with radioactive elements”, Paper accepted for presentation at the *International Biohydrometallurgy Symposium*, Madrid, June 20-23, 1999.

Groudeva, V.I. I.A. Ivanova, S.N. Groudev and G.C. Uzunov. 1993. "Enhanced oil recovery by stimulating the activity of the indigenous microflora of oil reservoirs". In A.E. Torma, M.L. Appel and C.L. Brierley (Eds.), *Biohydrometallurgical Technologies*, vol. II, pp. 349-356. TMS Minerals. Metals & Materials Society, Warrendale, PA.

FIELD-SCALE EVALUATION OF MICROBIAL MATS FOR METAL REMOVAL IN A FUNNEL-AND-GATE CONFIGURATION

Peter Phillips (Clark Atlanta University, Atlanta, Georgia)
Judith Bender (Clark Atlanta University, Atlanta, Georgia)
Crisandel F. Thornton (Clark Atlanta University, Atlanta, Georgia)

ABSTRACT: Microbial mat was pilot field tested for removal of an array of metals from an aqueous waste stream in a funnel-and-gate configuration under photosynthetic and darkened conditions. The gate contained microbial mat immobilized with inorganic components. The two recycled water test systems each contained 4,000 L into which 1 mg/L of cesium (Cs), chromium(VI) (Cr) and strontium (Sr) were spiked four times and barium (Ba), cobalt (Co) and lead (Pb) were spiked five times in 167 days. Flow rate was 1 L/min. Best metal removal by the microbial mat pads was Co > Pb > Ba > Sr > Cr = Cs. A total of 68.23 grams of metal were removed from the photosynthetic system (representing 63.5% of the total amount spiked) and a total of 67.36 grams were removed from the darkened system (representing 62.7% of the total amount spiked). In both systems redox ranged from -87.9 to -99.8 and pH from 8.2 to 8.7. Grab samples of microbial mat/polyester fiber substrate showed high metal concentrations (mg/kg), ranging from 2,100 and 1,500 mg/kg for Co to 30 and 85 mg/kg for Cr(VI) in the photosynthetic and darkened gates, respectively. With a TCLP analysis, the effluent showed all RCRA regulated metals below limits of detection.

INTRODUCTION

Biologically-reactive permeable barriers for oxidation and reduction, precipitation and stabilization of metals in groundwater may provide a low-cost solution to migrating metal-contaminated groundwater plumes. Iron-filled cells have been tested and implemented in different North American locations for reductive dechlorination of organic compounds (Focht et al., 1996; Gillham, 1996), but barrier cell contents which would function in metal removal have not been well researched. Microbial mats integrated into a funnel-and-gate type cell may be effective for metal removal from a groundwater plume.

Technological Development of Microbial Mat System. Microbial mats are consortia of microorganisms dominated by photosynthetic cyanobacteria. They also contain heterotrophic microbial species such as sulfate-reducing bacteria. Although mats function photosynthetically, this same microbial community offers a number of dark-function mechanisms, which are directly relevant to the remediation of specific contaminants in a below ground setting. Microbial mats can accumulate and tolerate high levels of toxic metals, metalloids and radionuclides.

The microbial mat system is in a technology transfer stage as a remediation technology (Phillips and Bender, 1995). In field pilot projects for the US Bureau of

Mines (Colorado) (Phillips et al., 1996) and the Tennessee Valley Authority (Alabama) (Phillips and Bender, 1998), it has been used to remove manganese, zinc and other heavy metals (silver, chromium, cadmium, copper, lead, nickel and iron) from mine drainage. At DOE's Oak Ridge National Laboratory, a pilot barrel bioreactor demonstrated the removal of uranium and nitrate from groundwater (SAIC unpublished report, 1998). The system has been used to removed low-level radioactive metals from wastewater of two nuclear power plants.

Objective. The microbial mat system was used to determine its capacity to remove an array of six metals from flowing water. The metals were barium (Ba), cesium (Cs), cobalt (Co), chromium(VI) (Cr), lead (Pb) and strontium (Sr).

MATERIALS AND METHODS

Live microbial mats integrating inorganic component supplements were grown on synthetic pads and configured into a permeable barrier of two separate simulations of funnel-and-gate type systems. One system was photosynthetic and one was darkened. The barriers intercepted flowing water from the supply tanks. Water passing through the barriers was collected in reservoirs and pumped back to the supply tanks. Metals were periodically spiked into the supply tanks. Metal concentrations were monitored throughout the two systems by ICP/MS analysis of water samples.

Field Pilot Test Facility. The pilot test facility was located at the Skidaway Institute of Oceanography's Bioremediation and Environmental Research Mesocosm, Savannah, Georgia. Two separate 4,000-L systems were constructed, one for testing a photosynthetic microbial mat gate and the other a darkened microbial mat gate. Microbial mat integrated with inorganic components was cultured on 4.6 m^2 of polyester fiber (PF-2 fiber, Aquatic Eco-Systems, Inc., Apopka, FL), cut into pieces and vertically hung in each gate parallel to the flow of water. The gates were located in the forward area of the funnel. Flow rate was 1 L/min by gravity from the supply tank into the funnel. Drainage from the funnel-and-gate was collected in a containment reservoir equipped with a sump pump which directed water back to the supply tank. All tanks were covered to inhibit photosynthesis and rainwater dilution in the supply and reservoir tanks and the funnel areas.

System Parameters. Metals tested were Ba, Cs, Co, Cr(VI), Pb and Sr. At spiking and periodic re-spiking, each metal concentration was brought up to 1 mg/L in the supply tank. Cs, Cr(VI) and Sr were spiked four times and Ba, Co and Pb were spiked five times during the 167-day period from September 1997 to March 1998. Sampling points were: the supply tank, the funnel (before the gate), the gate outflow and the containment reservoir. The water was recycled continuously from the reservoir to the supply tanks

RESULTS AND DISCUSSION

The order of greatest metal removal by the microbial mat pads was Co > Pb > Ba > Sr > Cr = Cs. A total of 68.23 grams of metal were removed from the photosynthetic system (representing 63.5% of the total amount spiked) and a total of 67.36 grams were removed from the darkened system (representing 62.7% of the total amount spiked). Except for Pb, metal removal diminished over time. This indicated that the metal ion binding sites may have been saturated or there was increased circuiting of water through the gate over time. Additionally, with a metal such as Cr(VI), which may require bacterial production of reductases to generate Cr(III), effective microbial populations may decrease over time.

During the operation of the systems, redox and pH, respectively, ranged from -87.9 to -99.8 and 8.2 to 8.7 in the photosynthetic system, and -88.7 to -97.7 and 8.2 to 8.7 in the darkened system.

Performance Results of Metal Removal. Relevant data for analysis covers 167 days of that period, from September 1997 - March 1998. Table 1, percentage removal of each metal in either the photosynthetic or darkened system, shows the relative removal efficiencies per metal. Tables 2 and 3 show gram weight removal of each metal, individually and combined, for the entire study period. Based on a 4.6 m^2 mat surface area, 14.8 g/m^2 of combined metals were removed in the photosynthetic system and 14.6 g/m^2 were removed in the darkened system.

TABLE 1. Percentage of metal removal in the photosynthetic and darkened gates during three time intervals. Cs, Cr(VI) and Sr were spiked 4 times and Ba, Co and Pb were spiked 5 times.

Photosynthetic Gate	Cs	Co	Sr	Cr	Pb	Ba
09/26/97-10/13/97	56.6	90.2	36.0	20.5	98.9	72.3
12/12/97-01/16/98	28.2	85.7	20.9	19.7	95.8	50.1
01/19/98-03/11/98	23.5	82.8	24.2	22.9	96.9	49.9
Darkened Gate	**Cs**	**Co**	**Sr**	**Cr**	**Pb**	**Ba**
09/26/97-10/13/97	58.6	90.1	38.6	39.6	99.4	74.9
12/12/97-01/16/98	15.6	79.7	9.1	6.9	82.3	39.3
01/19/98-03/11/98	18.0	79.3	15.5	17.0	93.4	47.8

Removal of Cs, which requires the inorganic component for removal, drops over time likely due to decreasing availability of sites for metal adsorption. Co and Pb removals are high and Sr and Cr(VI) are lowest. At the field site, calcium levels are high in the groundwater. Calcium directly competes with Sr for adsorption sites (H. Windom, Skidaway Institute director, personal communication). For metals, such as Cr(VI), which require an initial reduction to Cr(III) and subsequent precipitation as a hydroxide, bacterial metabolism is involved in the removal process, for example through the production of reductases by sulfate-reducing bacteria. Sulfide

TABLE 2. Total grams of metal removed in the photosynthetic and darkened gates during 167 days. Cs, Cr(VI) and Sr were spiked 4 times and Ba, Co and Pb were spiked 5 times.

Photosynthetic Gate	Cs	Co	Sr	Cr	Pb	Ba
Grams spiked	15.96	19.75	15.96	15.84	20.15	19.80
Grams removed	>9.29	15.15	9.81	8.43	13.19	12.36
Percent removal*		77	61	53	65	62
Darkened Gate	**Cs**	**Co**	**Sr**	**Cr**	**Pb**	**Ba**
Grams spiked	15.96	19.75	15.96	15.84	20.15	19.80
Grams removed	>8.54	16.50	9.18	5.52	15.60	12.02
Percent removal*		84	58	35	77	61

*Cesium data was misplaced for those dates between 11/10 - 12/08/98. Therefore, precise results were not computed.

TABLE 3. Total grams of metal removed during 167 days.

	Total metals spiked, g	Total metals removed, g	Percent removed
Photosynthetic system	107.46	68.23	63.5
Darkened system	107.46	67.36	62.7

generated by sulfate reducers metabolizing residual silage and dead algal cells, produce a sulfide pool which is important for precipitation of Pb. Effective microbial populations may decrease over time and thus account for decreasing metal removal capacity of the two systems. Another microbial factor that is significant in metal removal is the presence of bioflocculants or biopolymers. Measurements taken during this study showed high levels of bioflocculants (1/OD = 0/939). Bioflocculant has been demonstrated to be a very significant parameter in metal sequester in our laboratory research (Bender et al., 1994) for several of the metals used in this study.

Toxicity Characteristic Leaching Procedure (TCLP) Results. Upon disassembling the two gates, a grab sample of microbial mat/polyester fiber showed the mg metal concentration/kg substrate in Table 4. Both samples were submitted to a TCLP and the effluent showed all RCRA regulated metals in the sample leachate below limits of detection.

TABLE 4. TCLP for a discrete sample of microbial mat/polyester fiber from each gate.

	Photosynthetic Gate			Darkened Gate	
	Total Metals, mg/kg	TCLP Metals, mg/L	Regulatory Limit, mg/L	Total Metals, mg/kg	TCLP Metals, mg/L
Ba	810	15	100.0	420	13
Ce	300	11		140	8.4
Cr(VI)	30	<DL	5.0	85	<DL
Co	2,100	32		1,500	16
Pb	710	<DL	5.0	1,300	<DL
Sr	540	24		300	23

CONCLUSIONS

Real world contamination is seldom comprised of one single organic compound or metal. Generally, contaminated water, soils and sediments contain mixtures of these. Numerous microbially-based biotreatment systems will adsorb, precipitate or otherwise transform metals. Likewise, the degradation, or even complete mineralization, of organic compounds is common among microorganisms, such as fungi, bacteria and cyanobacteria. To achieve the two processes simultaneously, that of metal removal as well as organic degradation in one biological system, is unusual. Generally, the heavy metal or radioactive metal is toxic to the enzymatic systems that function in biodegradation of the organic material. Microbial mats have been shown to accomplish both (Phillips and Bender, 1995).

Bioremediation systems, designed for *in situ* treatment of recalcitrant compounds should ideally present several characteristics which are found in microbial mats: (1) ability to deal with mixed wastes; (2) contain a consortium of microbes which spontaneously generate heterogenous microzones (oxic/anoxic) in close proximity, thereby providing a variety of chemical environments; (3) degrading microbes that can penetrate soils and sediments; (4) mechanisms which transport contaminated materials from the sediment into the region of high microbial density; and (5) permanent immobilization of metals in the treatment matrix.

In this study, separate photosynthetic and darkened systems were tested. The rationale was that with light, microbial mats would create a very strong redox potential and support continuous generation of biomass to support sulfate-reducing bacteria. The question of a dark environment for photosynthetic organisms, such as cyanobacteria, is an important issue for underground *in situ* treatment. The provision of light to an underground treatment cell may be feasible depending upon the depth of the groundwater to be treated. Nevertheless, this pilot field study showed that microbial mats remained live and active in the dark as evidenced by a metal removal similar to that of the photosynthetic system and by the presence of metal-binding bioflocculants during 167 days.

ACKNOWLEDGMENTS

Support of the U.S. Department of Energy, Savannah River Site Grant Award No. DE-FG09-95SR18553 to Clark Atlanta University is gratefully acknowledged.

REFERENCES

Bender, J., S. Rodríguez-Eaton, U.M. Ekanemesang, and P. Phillips. 1994. "Characterization of Metal-Binding Bioflocculants Produced by the Cyanobacterial Component of Mixed Microbial Mats." *Applied and Environmental Microbiology 60*(7): 2311-2315.

Focht, R., J. Vogan, and S. O'Hannesin. 1996. "Field Application of Reactive Iron Walls for In-Situ Degradation of Volatile Organic Compounds in Groundwater." *Remediation/Summer 1996*: 81-93.

Gillham, R.W. 1996. "In Situ Treatment of Groundwater: Metal-Enhanced Degradation of Chlorinated Organic Contaminants." In M.M. Aral (Ed.), *Advances in Groundwater Pollution Control and Remediation*, pp. 249-274. Kluwer Academic Publishers, Netherlands.

Phillips, P., and J. Bender. 1998. "Bioremediation of Metals in Acid Tailings by Mixed Microbial Mats." In W. Geller, H. Klapper and W. Salomons (Eds.), *Acidic Mining Lakes, Acid Mine Drainage, Limnology and Reclamation*, pp. 347-363. Springer-Verlag, Berlin, Germany.

Phillips, P., J. Bender, S. Rodríguez-Eaton, J. Gould, and V. Shea-Albin.. 1996. "Removal of Metals from Mine Drainage by Microbial Mats." *Proceedings of the Fourth International Symposium on Environmental Issues and Waste Management in Energy and Mineral Production*, Volume 1: 475-482.

Phillips, P., and J. Bender. 1995. "Biological Remediation of Mixed Waste by Microbial Mats." *Federal Facilities Environmental Journal Autumn*: 77-85.

LONG-TERM STABILITY OF URANINITE FOLLOWING IN SITU REMEDIATION OF U(VI)

Abdesselam Abdelouas
Advanced Materials Laboratory, Center for Radioactive Waste Management, 1001 University Blvd., SE, Albuquerque, NM 87106
Werner Lutze and Eric H. Nuttall
Chemical and Nuclear Engineering, University of New Mexico

ABSTRACT: On a laboratory scale, we investigated biological reduction of uranium in groundwater. Indigenous bacteria were activated by adding ethanol and trimetaphosphate to groundwater in batch experiments. Uranium was reduced within few weeks. Poorly crystalline hydrated uraninite, $UO_2{\cdot}nH_2O$, precipitated together with mackinawite, $FeS_{0.9}$, because Fe(III) from the soil and sulfate in the water were reduced together with U(VI). Leaching experiments of uraninite in the presence of mackinawite using uncontaminated groundwater showed that the fraction of oxidized uraninite increases with increasing oxygen supply and decreases with increasing the amount of mackinawite precipitated on the soil. Most of the oxygen supplied by groundwater was consumed by mackinawite oxidation, thus ensuring the long-term stability of uraninite.

INTRODUCTION

In situ bioremediation of radioactive and non radioactive heavy metals is currently studied to clean up groundwaters and soils. Compared to the conventional technique "pump and treat" in situ bioremediation is more effective and less expensive according to recent studies (Quinton et al., 1997). Pump and treat is less suitable to remediate groundwaters contaminated with uranium because pumping the water may change the uranium speciation in solution and cause its partial sorption on the host rock minerals such as iron oxyhydroxides (Hsi and Langmuir, 1985). With in situ bioremediation both dissolved and adsorbed uranium can be reduced and precipitated as highly insoluble uraninite (Abdelouas et al., 1998).

Several microorganisms including iron and sulfate-reducing bacteria are able to reduce U(VI) to U(IV) and precipitate uraninite (Lovley, 1995). Fe-reducing bacteria such as *Shewanella putrefaciens* are capable of using U(VI) in place of Fe(III) as an electron acceptor to support cell growth (Lovley, 1991). Several sulfate-reducing bacteria including *Desulfovibrio* species are able to reduce U(VI) but can not grow by using U(VI) as the only electron acceptor (Lovley and Phillips, 1992, Lovley et al., 1993). These bacteria require sulfate in solution in order to grow and reduce U(VI). Iron and sulfate-reducing bacteria and heterotrophic bacteria in general are important groups for the subsurface e.g., groundwaters and soils (Bachofen et al., 1998). These microorganisms can be readily activated by addition of organic carbon and phosphate sources to the groundwaters (Abdelouas et al., 1999a).

Since the early 1990's research on U(VI) bioremediation in groundwater and soil intensified (e.g., Lovley et al., 1991, Gorby and Lovley, 1992, Lovley and Phillips, 1992, Phillips et al., 1995, Barton et al., 1996, Abdelouas et al., 1998, 1999b). These authors showed that low (<1 mg/L) or high (1.9 g/L) U(VI)-concentrations can be reduced by the enzymatic activity of Fe- or sulfate-reducing bacteria. The reduced uranium precipitates as poorly crystalline hydrated uraninite. Furthermore, the authors proposed uranium bioremediation as a potential technique for groundwater and soil clean up. If in situ bioremediation is considered rather than surface treatment, uraninite precipitates on the soil particles and remains underground. In this case, uraninite reoxidation and U(VI) migration with groundwater must be considered and investigated. Uraninite can be easily oxidized in contact with oxygen-rich groundwaters (Yanase et al., 1995). The only work that addressed the long-term stability of uraninite following bioprecipitation of uranium was conducted by Abdelouas et al. (1999b). The authors showed that the oxidation of uraninite in the presence of mackinawite is insignificant.

Objective. The objective of this study was to investigate the oxidation of uraninite precipitated enzymatically by sulfate-reducing bacteria in the presence of increasing amounts of dissolved oxygen and the mineral mackinawite.

MATERIALS AND METHODS

Groundwater and soil samples were collected from a uranium mill tailings site near Tuba City, Arizona, and from a former farm site located in the South Valley, Albuquerque, New Mexico. The Tuba City site groundwater is contaminated with U (0.25 mg/L), NO_3^- (1.2 g/L) and SO_4^{2-} (1.8 g/L); pH = 6.6. The farm site groundwater is contaminated with NO_3^- (464 mg/L) and contains 105 mg/L SO_4^{2-} with pH = 7.5. In the laboratory uranium was added to this water to reach a concentration of 3.7 mg/L.

Batch experiments using groundwater and soil were conducted in 160 mL serum bottles at room temperature (≈ 24°C). Experiments were conducted in triplicate and each experiment used 100 mL of groundwater and 8 g of autoclaved soil. Groundwater was amended with ethanol and sodium trimetaphosphate, $Na_3P_3O_9$. Ethanol was added to reach a stoichiometric ratio ethanol/sulfate of 2/3 according to reaction 1.

$$3\,SO_4^{2-} + 2\,C_2H_5OH \rightarrow 4\,HCO_3^- + 3\,HS^- + H^+ + 2\,H_2O \qquad (1)$$

Sodium trimetaphosphate was added to reach a final concentration of PO_4^{2-} of about 20 mg/L (Abdelouas et al., 1998). To assess the effect of sulfate and iron concentration on uranium reduction, we added, in some experiments, $FeSO_4{\cdot}7H_2O$ to the groundwater to reach an iron concentration up to 560 mg/L and a sulfate concentration up to 1.1 g/L.

At the end of the enzymatic reduction of U(VI), the water was removed and replaced by oxygen-rich uncontaminated groundwater from the Tuba City site (7 mg/L O_2) containing 12 mg/L NO_3^- and 10 mg/L SO_4^{2-} and with pH = 7.7, to study the reoxidation of uraninite. All manipulations were conducted under anoxic conditions in a glove box filled with nitrogen gas to prevent contamination with atmospheric oxygen.

Prior to analysis, groundwater samples were passed through a syringe filter with 0.2 μm pore size to remove biomass and mineral particles of the soil. Uranium was analyzed using a laser fluorescence analyzer (Scintrex UA-3) with a detection limit of 0.5 μg/L and a precision of ±15%. Sulfate were measured by liquid chromatography using a Dionex (DX-500) Ion Chromatograph with a precision of ±5% and a detection limit of 0.1 mg/L.

RESULTS AND DISCUSSION

Uranium reduction. An example of the results of experiments with contaminated groundwater from the mill tailings site in Tuba City is given in Figure 1, which shows the decrease of U(VI) and SO_4^{2-} concentrations over time. The activity of sulfate-reducing bacteria is confirmed by the decrease of sulfate concentration and the production of H_2S. The reduced uranium precipitates as poorly crystalline hydrated uraninite (Abdelouas et al., 1998). A fraction of sulfide precipitates as mackinawite, $FeS_{0.9}$ (Abdelouas et al., 1999b). The experiments using groundwater from the farm site showed that $FeSO_4.7H_2O$ enhanced the reduction of U(VI). The activity of sulfate-reducing bacteria was stimulated by the added sulfate (1.1 g/L), given that ethanol and phosphate are not limiting factors.

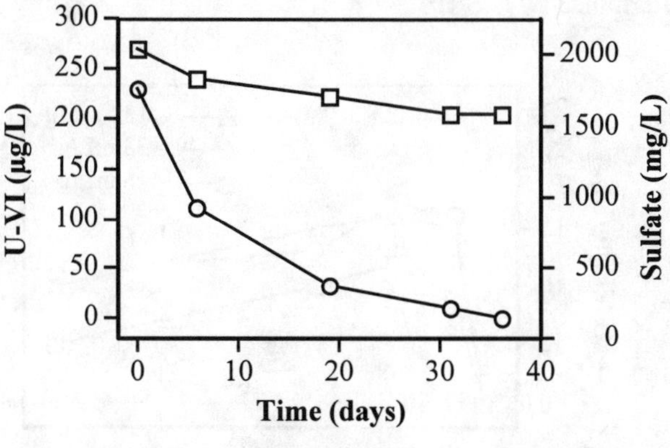

FIGURE 1. U(VI) and sulfate reduction at 24°C in contaminated groundwater from the mill tailings site (Tuba City, AZ) amended with ethanol and trimetaphosphate.

Uraninite reoxidation: effect of oxygen concentration in solution. The results of experiments in which 1 g of soil containing 3.1 μg of U(IV) was exposed to various volumes of oxygenated groundwater are given in Figure 2. Uraninite oxidation was determined by measuring the U(VI) concentration in solution (reaction 2).

$$UO_2 + 1/2\,O_2 + 2\,HCO_3^- \rightarrow UO_2(CO_3)_2^{2-} + H_2O \quad (2)$$

The oxygen mass supplied to 1 g of soil was 0.92 mg, 0.23 mg, and 0.12 mg in experiments using 132 mL, 33 mL, and 16 mL of uncontaminated groundwater, respectively (Figure 2). The figure shows that uraninite oxidation is fast and that the U(VI) concentration decreased with time. In all experiments enough oxygen was supplied to oxidize the total 3.1 μg of U(IV) but only a fraction was converted to U(VI). We measured sulfate concentration in the aqueous phase resulting from the oxidation of mackinawite (reaction 3).

$$FeS + 9/4\,O_2 + 3/2\,H_2O \rightarrow SO_4^{2-} + FeOOH + 2\,H^+ \quad (3)$$

The analytical results showed that according to reaction 3 more than 90% of the oxygen supplied by groundwater was consumed by mackinawite oxidation; and according to equation 2 only a small fraction of oxygen, less than 1%, was consumed by uraninite oxidation, the remaining oxygen was attributed to the oxidation of biomass generated during denitrification and sulfate reduction (Abdelouas et al., 1999b). The preferential oxidation of mackinawite relative to uraninite is expected because the redox intensity in natural water of sulfate, $p\varepsilon° = -3.75$, is lower than that of U(IV), $p\varepsilon° = +4.9$. The slow decrease of U(VI) concentration over time (Figure 2) is probably due to reduction and reprecipitation of uraninite because, after consumption of oxygen, the redox intensity of the solution is determined by H_2S/HS^-.

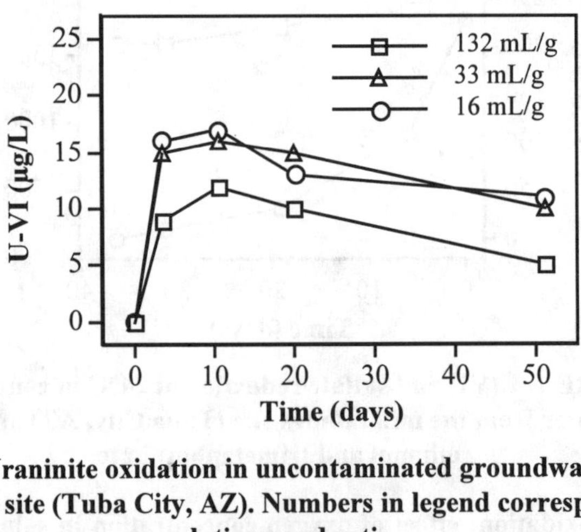

FIGURE 2. Uraninite oxidation in uncontaminated groundwater from the mill tailings site (Tuba City, AZ). Numbers in legend correspond to the solution:sand ratio.

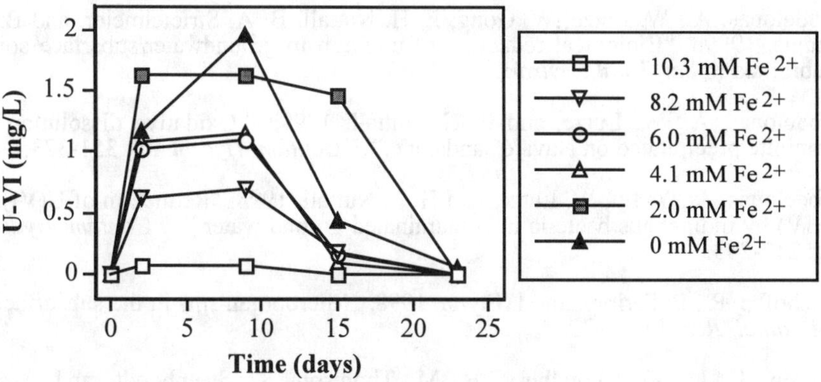

FIGURE 3. Uraninite oxidation in groundwater. Numbers in legend correspond to the Fe^{2+} added as $FeSO_4.7H_2O$ to the groundwater prior to uraninite precipitation.

Uraninite reoxidation: effect of mackinawite concentration. The results of uraninite oxidation using 8 g of soil containing 0.33 mg of U(IV) precipitated from the farm site groundwater are given in Figure 3. The amount of oxygen supplied in these experiments is constant (0.7 mg O_2) but the amount of mackinawite varied between about 0 and 1 mM. The amount of mackinawite was derived from the concentration of Fe^{2+} in solution prior to U(VI) reduction. Figure 3 shows that U(IV) is oxidized to U(VI) whose concentration reaches a maximum in all experiments and decreases to below 20 μg/L after 23 d. The decrease of U(VI) concentration is probably due to its reduction back to U(IV) after oxygen consumption by mackinawite oxidation. The amount of oxidized U(IV) decreases with increasing the amount of precipitated mackinawite. The details of the protective effect of mackinawite on uraninite are still under investigation.

CONCLUSIONS

This work shows that U(VI) can be removed from groundwater and precipitated as uraninite, $UO_2{\cdot}nH_2O$, via enzymatic reduction involving sulfate-reducing bacteria. A fraction of Fe(III) from the host rock is reduced to Fe(II) and precipitates as mackinawite, $FeS_{0.9}$. Oxidation of mackinawite consumes most of the oxygen supplied by groundwater leaving only a small fraction for uraninite oxidation. This fraction decreases with increasing ratio of Fe/U. If Fe(III) and sulfate are lacking in groundwater and soil, addition of sufficient iron and sulfate salts can help precipitate enough iron sulfide to protect uraninite from oxidation.

ACKNOWLEDGMENT

The research on which this paper is based was financed in part by the U.S. Department of Energy through the New Mexico Waste-management Education and Research Consortium (WERC).

REFERENCES

Abdelouas, A., W. Lutze, W. Gong, E. H. Nuttall, B. A. Strietelmeier, and B. J. Travis. 1999a. "Biological reduction of uranium in groundwater/subsrface soil." Submitted to *Sci. Total Environ.*

Abdelouas, A., W. Lutze, and E. H. Nuttall. 1999b. "Oxidative dissolution of uraninite precipitated on Navajo sandstone." *J. Contam. Hydrol. 36*: 351-373.

Abdelouas, A., Y. Lu, W. Lutze, and H. E. Nuttall. 1998. "Reduction of U(VI) to U(IV) by indigenous bacteria in contaminated ground water." *J. Contam. Hydrol. 35*: 217-233.

Bachofen, R., P. Ferloni, and I. Flynn. 1998. "Microorganisms in the subsurface." *Microbiol. Res. 153*: 1-22.

Barton, L. L., K. Choudhury, B. M. Thomson, K. Steenhoudt, and A. R. Groffman. 1996. "Bacterial reduction of soluble uranium: the first step of in situ immobilization of uranium." *Radio. Waste. Manag. Environ. Restor. 20*: 141-151.

Gorby, Y. A., and D. R. Lovley. 1992. "Enzymatic uranium precipitation." *Environ. Sci. Technol. 26*: 205-207.

Hsi, C-K. D., and D. Langmuir 1985. "Adsorption of uranyl onto ferric oxyhydroxides: Application of the surface complexation site-binding model." *Geochim. Cosmochim. Acta 49*: 1931-1941.

Lovley, D. R. 1991. "Dissimilatory Fe(III) and Mn(IV) reduction." *Microbiol. Rev. 55*: 259-287.

Lovley, D. R. 1995. *Microbial reduction of iron, manganese, and other metals*, In: L. Donald and L. Sparks (Editors), Advances in Agronomy, 54, Academic Press Inc., New York, 175-231.

Lovley, D. R., and E. J. P. Phillips. 1992. "Reduction of uranium by *Desulfovibrio desulfuricans*." *Appl. Environ. Microbiol. 58*: 850-856.

Lovley, D. R., E. E. Roden, E. J. P. Phillips, and J. C. Woodward. 1993. "Enzymatic iron and uranium reduction by sulfate-reducing bacteria." *Marine Geol. 113*: 41-53.

Lovley, D. R., E. J. P. Phillips, Y. A. Gorby, and E. R. Landa. 1991. Microbial reduction of uranium. *Nature 350*: 413-415.

Phillips, E. J. P., E. Landa, and D. R. Lovley. 1995. "Remediation of uranium contaminated soils with bicarbonate extraction and microbial U(VI) reduction." *J. Industr. Microbiol. 14*: 203-207.

Quinton, G. E., R. J. Buchanan, D. E. Ellis, and S. H. Shoemaker 1997. "A method to compare groundwater cleanup technologies." *Remediation*, Autumn, 7-16.

Yanase, N., T. Payne, and K. Sekine. 1995. "Groundwater geochemistry in the Koongarra ore deposit, Australia (I): Implications for uranium migration." *Geochem. J. 29*: 1-29.

CHROMATE REMOVAL FROM CONTAMINATED GROUNDWATER USING INDIGENOUS BACTERIA

J.S. McLean* and T.J. Beveridge (University of Guelph, Guelph, Ontario, Canada)
D. Phipps (Total Forest Industries, Guelph Ontario, Canada

ABSTRACT: Total Forest Industries Ltd., is a Canadian company, which owns a wood preserving business in Acton, Ontario. Over the many years of operation, the chromated copper-arsenate (CCA) solution used to preserve the wood has contaminated the underlying soils and groundwaters at the site. Copper and arsenic are bound to the soil, however due to unfavourable environmental conditions, chromium which is a highly toxic water-soluble metal in its hexavalent state, Cr(VI), has become mobile in the groundwaters and therefore demands immediate corrective action. After ceasing operations, the company is interested in using in situ or on-site bioremediation as a possible strategy to clean up the toxic metal contamination as opposed to using potentially hazardous chemical remediation alternatives. Recently, a bacterium capable of reducing chromium to an insoluble solid Cr(III) precipitate on its surface, thereby effectively removing the toxic Cr(VI) from solution, has been isolated from the contaminated site. An 16s rRNA sequence indicated that this efficient chromium reducer might be a yet unidentified gram-negative strain, which is tolerant to high levels of Cr(VI) (~500 mg/L) and, possibly, to Cu and As (40 mg/L). The exact mechanism by which this indigenous microorganism reduces aqueous Cr(VI) to a solid Cr(III) precipitate are varied since our results indicate there are a combination of biochemical and surface mediated reactions implicated in the process. The crude enzyme activity in a soluble cytoplasmic fraction of the cell yielded an estimated K_m of 23 mg/L and a V_{max} of 0.98 mgCr/h mg protein^{-1}.

INTRODUCTION

The discovery of microorganisms that will reduce chromium from the hexavalent to trivalent state may lead to applications in bioremediation field which are potentially more cost effective than current chemical remediation strategies. Conventional methods for removing toxic CrO_4^- include chemical reduction by lowering the pH, followed by precipitation. However, biological reduction should be more effective at neutral pH ranges, can be applied in-situ or ex-situ and require less energy input. Since the first reported *Pseudomonas* bacterial strains isolated from chromate-contaminated sewage sludge in the 1970's (Romanenko and Korenkov, 1977), several additional chromate reducing strains have been reported. These include other strains of *Pseudomonas*, as well as strains of *Micrococcus, Escherichia, Enterobacter, Bacillus, Aeromonas, Achromobacter, and Desulfomamaculum* species (Wang et al., 1989; Horitsu et al., 1987; Gvozdyak et al., 1986; Bopp and Ehrlich, 1988; Fude et al., 1994).

The overall objective of our study was to determine the potential of an indigenous bacterial strain to reduce and immobilize chromate in contaminated groundwater from a decommissioned wood-preservation site. This fundamental research into the mechanism and capacity of chromate reduction by a possibly novel isolate, designated CRB5, may lead to the development of an efficient and economically viable bioremediation strategy.

MATERIALS AND METHODS

Initial Sampling and Enrichment. Chromate resistant microorganisms were enriched from samples of CCA contaminated soil solutions and from the leachate storage pond on site. Viable colonies were selected from agar plates (aerobic and anaerobic) spiked with 10-60 mg/L chromium and enriched in minimal salts media broth containing 20 and 30 mg/L Cr(VI). An efficient chromate-reducing culture was isolated and purified by cycles of single colony isolation and liquid culture transfers in the presence of chromate.

Media. VB media was modified from Bopp et al.(1983), and was used in most of the experiments as a minimal salts media with additions of D-glucose CaCl and amino acids. VB broth was prepared by aseptically mixing 10 ml of VB concentrate, 10 ml of 25% D-glucose solution and 480 ml DW (all previously sterilized). VB concentrate contained the following (in grams per liter of distilled water): K_2HPO_4, 500; $Na(NH_4)HPO_4.4H_2O$, 175; citric acid, 100; and $MgSO_4$, 10. MEM 5X amino acids were added to some treatments 10mL/L as well as 20mL of 0.1M CaCl.

Batch Reactors. A 4.5 litre fermentation unit was used as a bioreactor for the batch experiments. Various parameters were measured over the time course of the reactions such as pH, O.D. (600nm), [Cr(VI)], [Cr_{tot}], and Eh. The reactor was inoculated with 1×10^9 cells/ml (OD_{600} =0.1).

Cell Fractionation and Anaerobic Metal Reduction Assay. Harvested cells were mechanically ruptured using a French press and the cellular fractions isolated by differential centrifugation in 0.01M HEPES buffer (pH=7). Supernatant fractions were obtained at 12,000xg (S_{12}), and 150,000xg (S_{150}). Whole cells were grown at 30°C and shaken in TSB for 24 hours. Harvested cells were washed 3X with the buffer solution. The tubes were inoculated with 1ml of culture to a concentration of approximately 1.5×10^7 cells/ml and grown anaerobically. Each treatment was conducted in duplicate with the initial sample taken at T=0 after the cells were inoculated.

Metal Analysis. Chromate was measured using the diphenylcarbazide method. Fe (II) production was assayed by reaction with ferrozine and the concentration determined colorimetrically. Co(III) was determined on a Dionex LC 50 liquid chromatograph. Uranium concentrations were determined colorimetrically. Protein was measured using the BCA assay. Eh readings were taken using a Corning combined pH-redox electrode.

Electron Microscopy. All electron micrographs were taken using a Philips EM400T operating at 100kV. Energy dispersive X-ray spectrometry was carried out using the EM400T which was coupled to a Link Analytical X-ray detector which allowed spectra to be collected over 100 s (live count time).

RESULTS

Enrichment and Isolation. In the initial enrichment process to isolate chromium tolerant bacteria, some mixed cultures grew rapidly and the solutions became turbid followed by a loss of the yellow color indicating they were transforming the chromate in solution. Further isolation and enrichment in minimal media containing chromate lead to the selection of two chromate reducing bacterial strains (CRBs) of which one was chosen for study (CRB5). CRB5 is a gram-negative motile rod which grows under aerobic or anaerobic conditions. A phylogenetic analysis using the 16s rRNA from the isolate resulted in a low homology with actinomycetes. The low homology indicated that this is most likely a novel species, which will require extensive research to fully characterize it physiology and biochemical attributes.

Chromate Reduction. The particular characteristics that would enable CRB5 to be used to reduce chromate in the groundwater or in an on-site bioreactor were tested. These included its ability to reduce chromate under aerobic and anaerobic conditions from the contaminated groundwater (Figure 1A). Once reduced to Cr^{3+}, chromium was then found to precipitate out on the surface of the cells as mixed amorphous Fe, Cr hydroxide mineral phases ($Fe_xCr_{1-x}(OH)_3$) (Figure 1B).

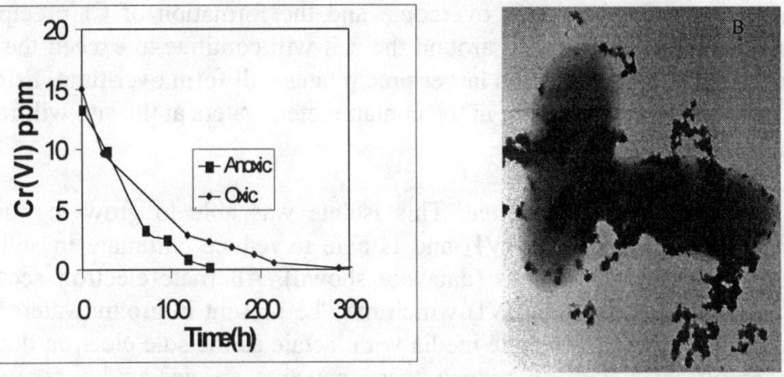

FIGURE 1. A) Cr(VI) reduction in the contaminated groundwater under aerobic and anaerobic conditions. B) Unstained electron micrograph of CRB5 and associated Cr precipitates after 22 hours incubation.

Chromium was also found within the cellular cytoplasm upon analysis of thin sections using TEM and EDS. In batch reactor experiments using VB amended with amino acids, CRB5 reduced chromate and lowered the redox potential from +200 to –400mV as well as decreasing the Ph from 6.8 to 6.2 (Figure 2A). The Eh decrease is a result of the bacterium using oxygen as a terminal electron receptor in the culture solution, which creates favourable geochemical conditions for the precipitation of $Cr(OH)_3$.

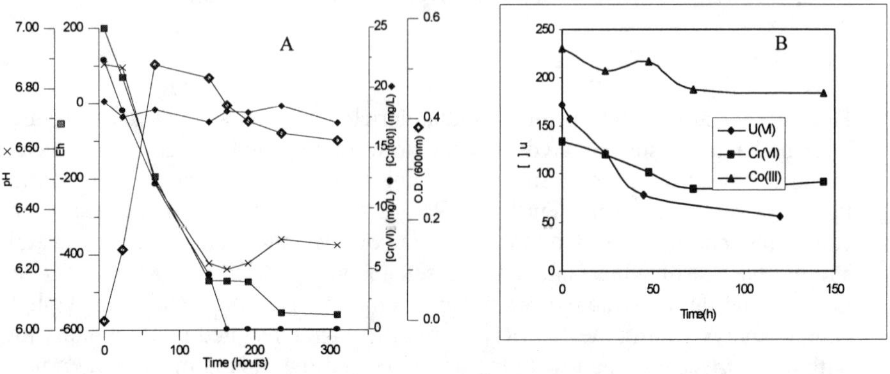

FIGURE 2.A) Cr(VI) reduction and associated changes in pH, Eh, O.D. in VBA media. B) eduction of Co(III), Cr(VI), and U(VI) under anaerobic conditions with lactate as an electon donor.

It seems likely that the initial reduction of Cr^{6+} to Cr^{3+} could be enzyme mediated (see below). Once the oxygen concentration is lowered, Cr^{6+} may serve as an alternate electron acceptor, reducing it to Cr^{3+} which is then free to bind stoichiometrically to the negatively charged surface functional groups on the cell surface. This is the first step in a two step mechanism proposed by Beveridge and Murray (1976), where the second step is a nucleation reaction that occurs when the activation energy barrier is overcome and the formation of Cr precipitates initiates. The microenvironment around the cell will continue to exceed the solubility product of $Cr(OH)_3$ (s) and larger precipitates will form over time. Effective removal of Cr(VI) from solution in the contaminated waters at the site will rely on this mechanism.

Metal Reduction and Tolerance. This isolate was able to grow in minimal media containing 500mg/L Cr(VI) and is able to reduce chromate in solutions containing 40 mg/L Cu and As (data not shown). Alternate electron acceptors such as Fe(III), Co(III), and U(VI) which may be present in groundwaters, were added separately to a bicarbonate media with lactate as the sole electron donor to test the ability of CRB5 to reduce these aqueous species under completely anaerobic conditions. It was found that CRB5 reduced Cr(VI), U(VI), and Co(III) whereas there was no reduction of Fe(III) (Figure2B).

Factors Affecting Reduction of Cr(VI). CRB5 was able to reduce Cr(VI) under low nutrient conditions (VB media) over a range of temperatures from 4 to 37°C and pH values from 4 to 9 (Figures 3A and 3B). The optimal conditions for reduction was 30°C and pH 9. As figure 3A shows, this isolate reduced chromate at temperatures less than 15°C, which are more likely in groundwater environments,

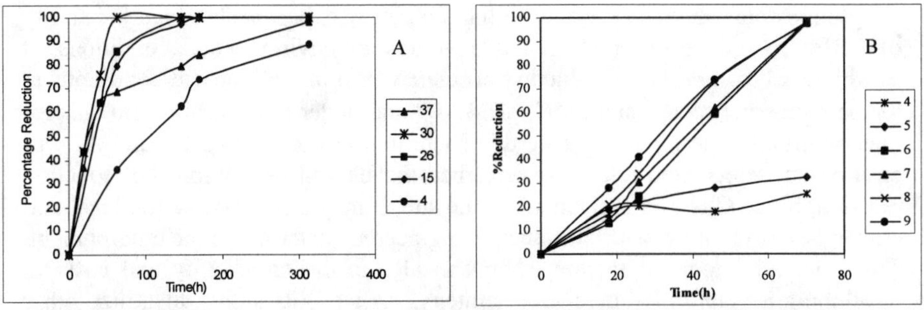

FIGURE 3. Percentage chromate reduction in VBA media (~10^7 cells inoculum) over time. A) Temperature effects on reduction rates. B) pH effects on reduction rates.

as well as at the pH of the groundwater at the site (pH=7.8). Nitrate appeared to have an inhibiting affect on chromium reduction possibly due to the fact that CRB5 is a strong denitrifier that completely reduces NO_3^- to N_2 gas.

Reduction Mechanism. After separating the cell membranes from the soluble components of the cell, the chromate reduction activity was found in the soluble fractions (S_{12},S_{150}) (Figure 4A). A kinetic analysis of the crude enzyme activity was conducted under anaerobic conditions in buffer only. The initial rate data was plotted in a Linweaver-Burk graph to estimate the Michaelis-Menton constant (K_m) and the maximum velocity (Figure 4B) which resulted in an apparent K_m of 23 mg/L (437μM) and a V_{max} of 0.98 mgCr/h mg protein^{-1} (19 μM/h mg protein^{-1}). These values are ten times as high as those reported for *Pseudomonas putida* PRS2000 (Ishibashi et al. 1990).

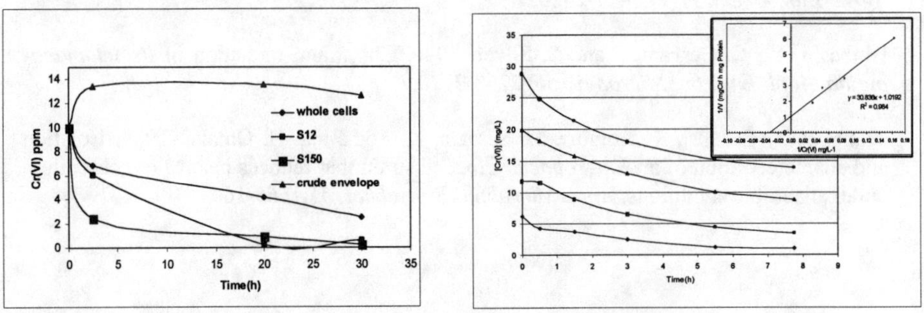

FIGURE 4. A) Anaerobic reduction of cell extracts B) Kinetics of anaerobic chromate reduction by cell free supernatent (S_{150}). Inset Linweaver-Burk plot.

CONCLUSIONS

Overall, several key microbial characteristics of this unclassified indigenous strain make it an ideal candidate for in situ use to reduce and immobilize chromate from the groundwater plume at the site. Initial laboratory experiments tested the ability of CRB5 to reduce chromate under a variety of environmental conditions. It exhibited a broad range of reduction efficiency in minimal nutrient conditions at temperatures between 4 and 37°C, pH 4 to 9 and under both aerobic and anaerobic conditions. The high tolerance this bacterium has to Cr(VI), Cu and As will ensure that unpredicted release of adsorbed metals will not inhibit the reduction reaction. Since CRB5 is indigenous to the site, it may also alleviate the legal and ethical issues involved with introducing engineered bacteria into the environment. These are all important factors when considering the complexity and costs in developing a viable bioremediation strategy. As well, its ability to utilize other soluble metal species as terminal electron acceptors including radioactive U(VI), enhances its potential to remediate other metal contaminated areas.

REFERENCES

Beveridge,T.J. and R.G.E. Murray. 1976. Uptake and retention of metals by cell walls of *Bacillus subtilis*. *J.Bacteriol.* *127*:1502-1518.

Bopp, L.H., A.M. Chakrabarty, and H.L. Ehrlich. 1983. Chromate resistance plasmid in *Pseudomonas flourescens*. *J. Bacteriol.* *155*:1105-1109.

Fude, L., B. Harris, M.M. Urrutia, and T.J. Beveridge. 1994. Reduction of Cr(VI) by a consortium of sulfate-reducing bacteria. *Appl.Environ. Microbiol.* *60*:1525-1531.

Gvozdyak P.I., N.F. Mogilevich, A.F. Ryl'skii, and N.I. Grishchenko. 1986. Reduction of hexavalent chromium by collection strains of bacteria. *Mikrobiologiya* *55*:962-965.

Horitsu, H., S. Futo, Y. Miyazawa, S. Ogai, and K. Kawai. 1987. Enzymatic reduction of hexavalent chromium by hexavalent chromium tolerant *Pseudomonas ambigua* G-1. *Agric. Biol. Chem.* *51*(9):2417-2420.

Ishibashi, Y., C. Cervantes, and S. Silver. 1990. Chromium reduction in *Pseudomonas putida*. *Appl. Environ. Microbiol.* *56*: 2268-2270.

Wang, P.C., T. Mori, K. Komori, M. Sasatsu, K. Toda, and H. Ohtake. 1989. Isolation and characterization of an *Enterobacter cloacae* strain that reduces hexavalent chromium under anaerobic conditions. *Appl. Microbiol. Biotechnol.* *55*:1665-1669.

ENHANCED REMOVAL OF COPPER FROM CONTAMINATED SILT SOIL USING BIOELECTROKINESIS.

Giacomo Maini and Ajay K. Sharman (IBS Viridian Ltd., Whitstable, UK)
Simon A. Jackman, Christopher J. Knowles (Oxford Centre for Environmental Biotechnology, Oxford, UK)
Garry Sunderland (EA Technology Ltd., Capenhurst, UK)

ABSTRACT: An initial electrokinetic experiment was performed using a characterised silt soil, spiked with 1000 mg/kg of Cu. The results showed that 92% of the Cu was mobilised and recovered in the cathode solution after 15 days treatment. The pH in the soil decreased from 7.5 to 2.1. Time course analyses of soil cores indicated the mobilisation of the metals present in the soil followed the order Ca>Mn>Cu while SO_4^{2-}, Mg and Fe showed no relative mobility. Cu contaminated silt soil was further amended with 5% (w/w) elemental sulphur in order to stimulate the activity of Sulphur Oxidising Bacteria (SOBs). Soil kept at constant moisture (30%w/v) was acidified to pH 5.4 after 90 days incubation at room temperature (~20°C). Further acidification did not occur, as the presence of partially mobile Cu seemed to inhibit the SOB activity. The soil was packed into the electrokinetic reactor operated at the same current density used in the initial study (3.72 A/m^2). The data showed that after a 10 day treatment period the pH of the soil profile was <3.0 and 80 % of the Cu was successfully removed. During the electrokinetic process SO_4^{2-} concentration across the soil profile significantly increased suggesting that the toxic inhibition of the Cu upon the SOBs was removed as the metal was recovered at the cathode. Ca, Mn, Mg and Fe showed the same order of mobility as reported in the initial run while sulphate accumulated as a band in the middle of the soil cell. Although the total Cu removal (86%) after 16 days treatment was similar to the initial run without SOBs, a cost assessment based on the power consumption revealed that this integrated process could potentially produce substantial energy savings in a full scale treatment system.

INTRODUCTION

The principle of electrokinesis is the application of low voltage DC current causing a pH gradient and an electroosmotic flow of water, which facilitates the directional movement/recovery of heavy metals (Yeung and Datla, 1995). Acidification of soils is obtained by the electrolysis of water at the electrodes. Under oxidative conditions at the anode hydrogen ions [H^+] are produced. These positively charged ions will move through the soil pore water towards the cathode lowering the pH of the soil thereby mobilising the metals species. Similarly electrolysis of water at the cathode produces an excess of [OH^-] which can produce alkaline conditions with the consequent precipitation of heavy metals into insoluble forms. In order to avoid heavy metals precipitation, acidic conditions can be successfully maintained at the cathode by adding a buffering agent such as acetic acid, carbon dioxide and hydrochloric acid (Hicks and Tondorf, 1994). Acetic acid has the advantages of both keeping the cathode under acid conditions and to increase the solubility of certain metals such as Cu, Zn, Pb and Ca in the cathode compartment. Nevertheless, the electrokinetic process alone may be ineffective in removing metals in soil with high buffering capacity where high concentration of [H^+] are required before achieving any significant drop in pH through the soil profile (Reed *et al.*, 1996). Under these circumstances electrokinetic remediation may prove to be economically unacceptable (Will, 1995).

The presence of DC voltage in the soil has been shown to increase the activity of the indigenous bacterial population (Electorowicz and Boeva, 1996). Moreover, recent shake flask studies (Jackman *et al.,* 1999) have shown that DC field can stimulate the activity of the naturally occurring Sulphur Oxidising Bacteria (SOBs). These bacteria have the capacity to use reduced sulphur as an electron acceptor producing H_2SO_4 as the end product thereby 'naturally' acidifying the soil. Therefore, amending soil with sufficient elemental sulphur or sulphide can produce a soil acidification which may result in the 'bioleaching' of certain metal species (White *et al.,* 1998). The integration of bioleaching caused by the SOBs to the electrokinetic process may result in an enhanced removal of heavy metals from difficult soils with a reduction in the energy costs which are usually associated with the electrokinetic process. This paper investigates the potential advantages of using the integrated process combining the bioleaching of the SOBs with electrokinesis to a well-characterised silt soil spiked with 1000 mg/kg of Cu.

MATERIALS AND METHODS

A silt soil was used. The characteristics of the silt soil are presented in Table 1.

TABLE 1. Silt soil characteristics.

Parameter	Value
pH	8.1
Extractable phosphorus (mg/L)	24
Cationic exchange capacity (me/100g)	21.7
Soil buffering capacity (g/kg) *(calculated as the amount of [H^+] from sulphuric acid required to acidify the soil to pH 2)*	0.52
Particle size 2000-600 µm (%)	1
Particle size 600-212 µm (%)	4
Particle size 212-63 µm (%)	7
Particle size 63-20 µm (%)	43
Particle size 20-2 µm (%)	19
Particle size <2 µm (%)	26
Calcium carbonate (%)	2
Total Sulphur (%)	0.08
Total nitrogen (%)	0.25
Organic matter (5)	3.40
Organic carbon (%)	1.97
C:N ratio	7.9:1

Silt soil was spiked with a solution containing $Cu(NO_3)_2$ in order to provide a concentration of 1000 mg/kg as Cu. Approximately 1 kg of soil was tightly packed in the reactor prior to starting the experiment. A diagram of the electrokinetic rig is presented in figure. 1. The anode consisted of a carbon felt while the cathode was composed of a stainless steel electrode, housed in a semiporous membrane, which allowed the recirculation of the catholyte.

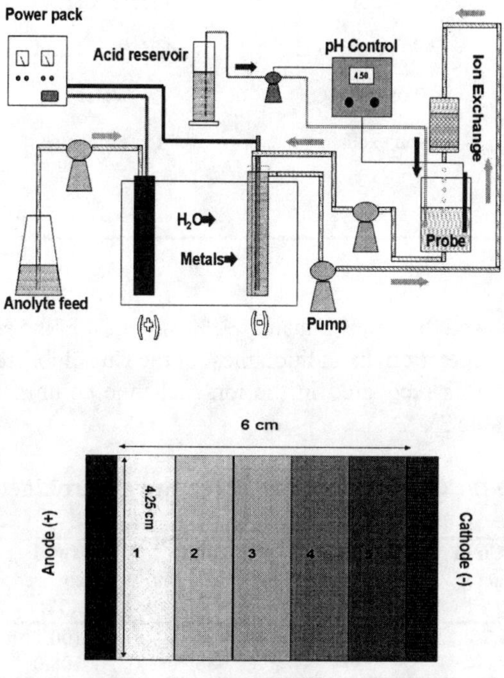

FIGURE 1. Electrokinetic reactor set up and cells sampling diagram.

The reactor vessel was constructed by six cells allowing a modular design for a time course sampling of the soil profile. This was achieved from further subdividing each cell into 5 subsections at 1 cm intervals with 2 sampling cores per section.

Indigenous SOBs present in the spiked silt soil were stimulated by the addition of 5% (w/w) elemental sulphur and soil was maintained at 30% (w/v) soil moisture. Soil was incubated at room temperature (~20°C) and mixed once a week in order to ensure the oxygen necessary for the oxidation process. Sulphate concentration in soil, pH and moisture were monitored weekly. Once the soil profile showed no further decrease in pH, the soil was transferred into the electrokinetic reactor in order to commence the electrokinetic experiment. The experimental conditions used during the two experiments are presented in Table 2.

Water was added at the anode to prevent the soil drying out during the electrokinetic runs. The pH control unit was set to maintain the catholyte at pH ≤ 4.5 using a concentrated acetic acid solution. Soil samples were taken from the electrokinetic reactor during the experiment at time intervals.

At the end of the experiment soil cores were taken from each cell and analysed for soil moisture, exchangeable sulphate and pH. Analyses of Cu, Ca, Fe, Mn and Mg in the soil cores were performed by X-ray fluorescence spectrometry (XRF). Concentrations of Cu, Ca, Mn, Mg and Fe in the ion exchange elution were measured using Atomic Absorption Spectrometry (AAS).

TABLE 2. Condition imposed to the two electrokinetic experiments.

Parameter	Run 1	Run 2
Soil Type	Silt soil	Silt + 5% w/w sulphur
Porosity	Approx. 10^{-5} m/s	approx. 10^{-5} m/s
Contamination	Cu 1000 mg/kg	Cu 1000 mg/kg
Current Density	3.72 A/m^2	3.72 A/m^2
Catholyte addition	4 M Acetic acid	4 M Acetic acid
Anolyte addition	0.05M H_2SO_4	Water
Metal recovery device @ the cathode	Ion exchange	Ion exchange
Time of run	15 days	16 days

RESULTS

The result of Run 1 showed that approximately 92% of the Cu in the silt soil was removed after 15 days operation. In addition most of the Cu solubilised and removed in the catholyte was recovered in the ion-exchange column as shown in the mass balance (Table 3).

TABLE 3. Calculation of the Cu mass balance in the two electrokinetic experiments.

Run	Initial Cu concentration (mg/kg)	Total Cu in the reactor (mg)	Final Cu concentration (mg/kg)	Total removed (%)	Total recovered (%)
1	1080	1086.76	89	92	100.1
2	920	901.35	137	86	108.0

The soil pH decreased from 7.5 to 2.8 during the 15 days electrokinetic run. The addition of [H^+] from the sulphuric acid added (~0.086 g/kg soil) at the anode was well below the buffering capacity of the soil, thus suggesting that the electrokinetic process alone can produce enough acid to decrease the pH (Table 4). The mobilisation of the metal present in soil was Ca>Mn>Cu and was affected by the decrease in the soil pH (Fig.2). In contrast, Fe and Mg were poorly mobilised and low concentrations were found to be dissolved in the catholyte. Metal banding across the soil cell during electrokinetic remediation was found to be caused by the change in the pH in the soil with the highest concentration measured in the proximity of the pH 'jump' (Table 4).

TABLE 4. Cu distribution, SO_4^{2-} and pH profile during electrokinetic remediation in Run 1. Highlighted values showed the greatest Cu and SO_4^{2-} concentrations in soil compared to the highest pH values.

Parameter	Time (d)	Cu removal (%)	Distance from the anode (cm) 1	2	3	4	5
Cu (mg/kg)			231	2323	1206	1166	1132
SO_4^{2-}(mg/kg)	2	0	2786	1464	0	0	0
pH			2.03	3.65	7.24	7.61	6.69
Cu (mg/kg)			73	105	313	466	906
SO_4^{2-}(mg/kg)	9	65	2909	2446	82	62	41
pH			2.38	2.61	3.27	3.35	4.08
Cu (mg/kg)			40	69	107	156	196
SO_4^{2-}(mg/kg)	15	92	2407	2174	3805	79	123
pH			2.12	2.40	2.89	3.32	3.42

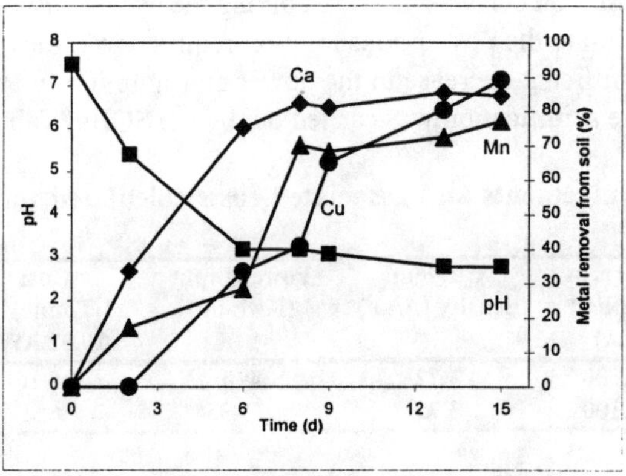

FIGURE 2. Removal of Cu and earth metals during electrokinetic remediation in relationship to the soil pH.

Silt soil amended with sulphur (5%, w/w) was packed into the electrokinetic reactor when it reached a pH value of 5.4 and an average sulphate level of 3101 mg/kg. Once the electrokinetic remediation was initiated more than 80% of Cu was removed after 10 days. This was partly caused by the lower initial pH of the silt soil (i.e. pH 5.4), which was already acidified by the SOBs. Again, most of the Cu removed from the soil was recovered in the ion-exchange column (Table 3) and Cu metal banding was observed in a similar manner as reported for Run 1 (Table 5).

The final removal of Cu from the soil after 16 days treatment was approximately 86%. Ca and Mn were found to be readily mobilised in the presence of the SOBs while Fe and Mg showed almost no movement. Sulphate appeared to accumulate as a band in the middle of the reactor cell (Table 5) and its average concentration across the cell increased significantly during the electrokinetic run. The increase in the average sulphate concentration at day 10 was 3.8 fold higher than at day 0. A decrease the sulphate concentration within the soil profile was recorded after day 14 as the negatively charged sulphate molecules started to migrate into the carbon anodic felt.

TABLE 5. Cu distribution, SO_4^{2-} and pH profile during electrokinetic remediation in Run 2. Highlighted values showed the greatest Cu and SO_4^{2-} concentrations in soil compared to the highest pH values.

Parameter	Time (d)	Cu removal (%)	Distance from the anode (cm) 1	2	3	4	5
Cu (mg/kg)			384	836	2680	1360	1160
SO_4^{2-}(mg/kg)	3	0	8610	11705	2871	4011	2136
pH			1.83	2.4	4.08	5.34	5.27
Cu (mg/kg)			103	124	168	225	310
SO_4^{2-}(mg/kg)	10	80	5493	17313	16635	19981	5483
pH			2.19	2.11	2.10	2.15	2.66
Cu (mg/kg)			103	100	132	149	202
SO_4^{2-} (mg/kg)	16	86	6335	7011	7399	4689	6876
pH			1.97	2.05	2.18	2.02	2.31

DC voltage across the soil cell was measured during the two electrokinetic runs; this was used to calculate the power consumption required to remove Cu from soil. There was a significant decrease in the power consumption in Run 2 where the initial step of the acidification was carried out by the SOBs (Table 6).

TABLE 6. Power requirements and associated costs calculated for the two electrokinetic runs

Run	Run time (Days)	Current applied (A)	Current density (A/m^2)	Power Input (Kwh/m^3)	Cost ($/Tonne)* @ $0.32 kWh
1	15	0.100	3.72	490	26.16
2	16	0.100	3.72	173.95	9.26

1 m^3 of soil is approx. 1.5 tonne

DISCUSSION

These experiments highlighted potential of electrokinesis alone and in combination with SOBs in remediating soils contaminated with heavy metals. The data of both Run 1 and 2 revealed that soil acidification must be achieved prior to mobilising Cu. Cu started to be removed once the soil had achieved a level < pH 4.0. Under these conditions Cu is present in the pore water as the ionic form Cu^{2+} and electromigrated towards the cathode. However, acidification of the soil could be achieved only when the carbonate buffering system could be overcome. The $CaCO_3$ in the silt soil was ~2% therefore large quantities of Ca were removed and recovered at the cathode. Ca did not precipitated at the cathode, as the solubility of calcium acetate is greater than 1g/L. Mn showed similar behaviour, while Mg showed almost no mobility. Fe is present in the silt soil at concentration of 30,000 mg/kg, but seemed to be locked up as insoluble ferric hydroxide species, which becomes mobile only at pH values < 3.0 (Ulrich, 1991).

In Run 2, it was assumed initially that some of the Ca^{2+} present in the pore water could react with the excess SO_4^{2-} produced by the SOBs and precipitate as insoluble $CaSO_4$, nevertheless the data did not corroborate this assumption.

Data for SO_4^{2-} were similar to those reported by Acar *et al.* (1997). SO_4^{2-} moved from the cathode toward the anode as result of electromigration, however this was in part counteracted by the electroosmotic flux of the water towards the cathode. For instance, during Run 1 the H_2SO_4 was added at the anode and the SO_4^{2-} remained confined near the anode although some migration caused by the water flux towards the cathode was recorded towards the end of the run. On the other hand, in Run 2 the SO_4^{2-} concentration was evenly distributed across the soil profile at T=0 as the result of SOBs acidification. As electrokinetic remediation proceeded a band in the middle of the soil cell was observed from day 3 due to the counteracting effects of electroosmosis and electromigration. This was also in combination with a general increase in SO_4^{2-} concentration caused by the SOBs activity. The electromigration effect has been reported to be more pronounced than electroosmosis (Probstein and Hicks, 1993) and by day 16 most of the $SO_4^{=}$ was confined near or within the anode.

The increase in the SO_4^{2-} production in the cell inoculated with SOBs indicated an enhanced microbial activity. Previous experiments conducted by Jackman *et al.*, 1999 in shaken flasks showed that indigenous SOBs in similar sulphur amended silt soil could not achieve acidification < pH 4.0, as the mobile Cu^{2+} inhibited any further bacterial activity. Similarly, pre-acidified soil did not reach acidification < pH 5.4 prior to starting the electrokinetic remediation, even after 90 days incubation. It is possible that the 'switch on' of the electrokinetic process, with the consequent migration of the Cu^{2+} towards the anode, removed the inhibition caused by this heavy metal upon the indigenous SOBs population. A previous study (Tuovinen *et al.*, 1971) has reported that high concentrations of ionic Cu^{2+} (i.e. >1000 mg/kg) may prolong the lag phase and partly inhibit the metabolism of *Thiobacilli.*

The electrokinetic cell inoculated with SOBs showed faster initial removal rates, however the overall level of removal was similar to the experiment without SOBs activity. Nevertheless, the electrokinetic process coupled with bioleaching showed significantly lower power requirements. These initial results suggest that an integrated electrokinetic and microbial bioleaching system (Bioelectrokinesis) may have substantial advantages over the single electrokinetic process. A detailed consideration of a number of operational parameters such as electrode costs, applicability to soil with high buffering capacity, presence of weathered metals will be required to determine the commercial viability of the process.

REFERENCES

Acar, Y. B., M. F. Rabbi, and E. E Ozsu. 1997 "Electrokinetic Injection of Ammonium and Sulphate Ions Into Sand and Kaolinite Beds." *Journal of Geotechnical and Geoenvironmental Engineering* 123:239-249

Electorowicz, M and V. Boeva. 1996. " Electrokinetic Supply of Nutrients in Soil Bioremediation." *Environmental Technology*. 17:1339-1349

Hicks. R.E. and S. Tondorf. 1994. "Electrorestoration of Metal Contaminated Soils." *Environ. Sci. Technol.* 28: 2203-2210

Jackman, S. A, G. Maini, A. K Sharman and C. J. Knowles. 1999 "The effects of Direct Electric Current Upon the Viability and Metabolism of Acidophilic Bacteria." *Enzyme Microbial Technology*. In press

Probstein, R. F. and R. E. Hicks. 1993 "Removal of Contaminants from Soil by Electric Fields." *Science*, 260: 498-503

Reed, B. E., P. C. Carriere, J.C. Thompson and J. H. Hatfield. 1996 " Electronic (EK) Remediation of a Contaminated Soil at Several Pb Concentrations and Applied Voltages." *Journal of Soil Contamination*. 5 (2): 95-120

Tuovinen, O. H., S. I. Niemelä and H. G. Gyllenberg. 1971 "Tolerance of *Thiobacillus ferrooxidans* to Some Metals" *Antonie van Leeuwenhoek,* 37: 489-496

Ulrich B. 1991 "Deposition of Acids and Metal Compounds." In E. Merian Eds. *Metals and Their Compounds in the Environment.* VCH Publishers, Inc., New York, N.Y.

Yeung, A. T. and S. Datla. 1995 "Fundamental Formulation of Electrokinetic Extraction of Contaminants from Soil" *Can. Geotech. J.* 32: 569-583

White, C., A. K. Sharman and G. M. Gadd. 1998 "An Integrated Microbial Process for the Bioremediation of Soil Contaminated with Heavy Metals." *Nature Biotechnology* 16: 572-575

Will, F. G. 1995 "Removing Toxic Substances from the Soil Using Electrochemistry." *Chemistry & Industry* 5:376-379

ORGANIC ELECTRON DONORS FOR THE MICROBIAL Cr(VI) REDUCTION

Jihua Hong and Guy W. Sewell (USEPA, Ada, OK, USA)

ABSTRACT: Sucrose-amended, yeast extract-amended, and lactate-amended systems were observed to be effective for microbial Cr(VI) reduction in an organic-carbon-poor subsurface environment. Approximately 10 mg/l of Cr(VI) was completely reduced in these systems during a 17-day experiment. In an additional experiment, nitrate, Cr(VI), and sulfate reduction were observed to depend on supplies of the electron donor (lactate). A reduction sequence of nitrate, Cr(VI), and sulfate was observed in the system. The reduction rate of the respiked Cr(VI) in the system was estimated to be 0.2203 h^{-1}.

INTRODUCTION

Microbial Cr(VI) reduction has been investigated in recent years (Chen and Hao, 1997; Chirwa and Wang, 1997; Krauter et al., 1996; Lovley and Pillips, 1994; Shen et al., 1996; Wang et al., 1989). The investigations were focused on microbial Cr(VI) reduction in fully-supplied-nutrition systems. Most were based on bacterial growth in pure culture, of species isolated from soils and other environmental materials, or from stock culture collections. Some papers have also reported Cr(VI) reduction under special conditions such as with the addition of nitrate and sulfate which are common in subsurface environments (Chen and Hao, 1996; Wang and Xiao, 1995). These investigations have shown that many bacteria in subsurface environments can contribute to Cr(VI) reduction. However, questions still remain:

- In a remediation action for Cr(VI)-contaminated sites, what are the most useful electron donors for the microbial Cr(VI) reduction?
- What are the critical or rate limiting factors for the microbial Cr(VI) reduction in the remediation?
- What is the redox range in subsurface systems of the microbial Cr(VI) reduction process?

In this study, five organic electron donors were used to determine which were suitable for microbial Cr(VI) reduction. The effects of multiple concentrations of lactate on Cr(VI) reduction were investigated. The fate of nitrate and sulfate in the Cr(VI) reduction test systems was investigated.

MATERIALS AND METHODS

Subsurface material was obtained from near a municipal landfill located in Norman, OK. The samples were stored at 4 °C in glass containers sealed with air-tight lids until use. An artificial groundwater medium used in the experiment was prepared by dissolving 800 mg of $(NH_4)_2HPO_4$, 200 mg of KH_2PO_4, 100 mg of NaCl, 200 mg of $MgSO_4.7H_2O$, and 100 mg of $CaCl_2.2H_2O$ in 1.0 L of distilled water.

Stock solutions of 0.5 M sucrose, 0.5 M sodium benzoate, 1 M of sodium acetate, 1 M sodium lactate, 0.4 M of K_2CrO_4, 1M KNO_3 and 10% yeast extract (w/v) were freshly prepared with DI water. The stock solutions were degassed and then autoclaved at 121 °C for 25 min and moved to the glovebox for future use.

The anaerobic glovebox (Coy) was filled with a nitrogen (92.5%)/hydrogen (7.5%) gas mix. The atmosphere of the glovebox was continuously re-circulated by two fan boxes through a palladium catalyst which in the presence of hydrogen quickly removed any trace oxygen. A Coy oxygen-hydrogen analyzer in the glovebox monitored oxygen concentrations during the experiment, and showed no detectable oxygen (< 1 ppm) was present during construction, sampling and incubation of the test systems.

Microcosms were prepared aseptically using sterile 1-L screw-cap amber bottles and were incubated at room temperature (22 °C), in the glovebox. All glassware, caps, tips and preparation supplies were autoclaved at 121 °C for 25 min and were stored in the glovebox before use to remove any traces of oxygen.

For the different electron donor-amended systems, aquifer solids were mixed and distributed to amber experimental bottles. 200 g of the wet material was added to each bottle. The bottles were then filled with 350 ml of artificial groundwater and 350 ml of Byird's Mill water, which is a locally produced groundwater. The artificial groundwater and the Byird's Mill water had been purged with N_2 for 20 min before use. The bottles were then moved into the glovebox. The systems designed as controls were purged with N_2, then autoclaved for 45 min at 121 °C on 2 successive days. Sterile stocks of electron donors, KNO_3 and Cr(VI) were then added to the bottles. The concentrations of sucrose, benzoate, acetate and lactate in the different systems were approximately 14 mM, while the yeast extract concentration was approximately 700 mg/l. These concentrations are equivalent to 1246 mg/l for lactate, 840 mg/l for acetate, 4792 mg/l for sucrose, and 1709 mg/l for benzoate, respectively. The concentrations of NO_3-N, Cr(VI) and SO_4^{2-} were approximately 12 mg/l, 10 mg/l and 40 mg/l, respectively.

Lactate, as a representative organic electron donor, was used to determine the effect of concentration on the microbial reduction of Cr(VI), nitrate and sulfate. Four different concentrations (0.1, 1, 3, and 10 mM) of lactate were applied to the test systems.

Routine sampling was accomplished by transferring 1 ml of suspension to a 1.5-ml plastic microcentrifuge tube. The sample was then centrifuged at 10,000 rpm for 15 min to remove any suspended solids. Cr(VI) was determined using a Thermo Separations capillary electrophoresis (CE) system equipped with a UV detector and a 70 cm x 75 μm i.d. capillary cassette (Model Spectra Phoresis 500, San Jose, CA). The separation was performed at 30 °C by applying a negative voltage of 25 kV. Cr(VI) was determined at 273 nm. Limit of quantification for Cr(VI) was 0.05 mg/l.

Organic acids, including lactic, formic, acetic, propionic and butyric acids, were determined with a Dionex 500 ion-chromatographic system (Dionex, Sunnyvale, CA), which was composed of a GP40 gradient pump, an ED40 electrochemical detector, a 50 μl PTFE sample loop and a S40 automatic sample loading device.

Anion concentrations (NO_3^-, SO_4^{2-}) were determined using a Quanta™ 4000 capillary ion analysis method. More details can be found elsewhere (Hong and Sewell, 1998).

RESULTS AND DISCUSSION

Comparison of electron-donor amended systems. From Table 1, we can see 100% of Cr(VI) in the sucrose-amended, lactate-amended, and yeast extracted-amended systems was reduced during a 17-day experiment, while Cr(VI) reduction in the acetate-amended and benzoate-amended systems were only 34.5% and 13.7%, respectively.

In the sucrose-amended system, pH decreased to below 5 during the

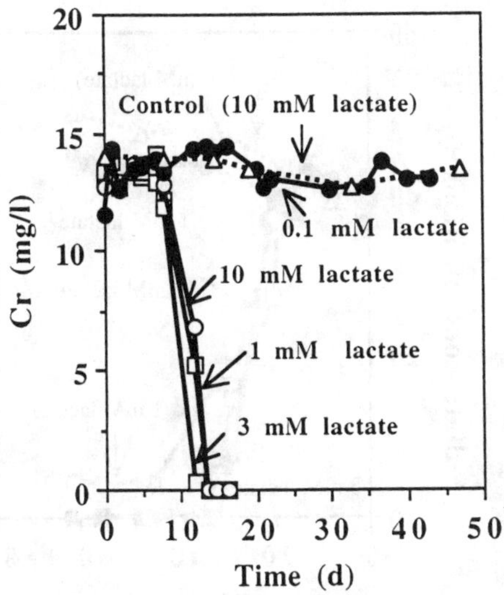

FIGURE 1. Effects of lactate concentrations on Cr(VI) reduction

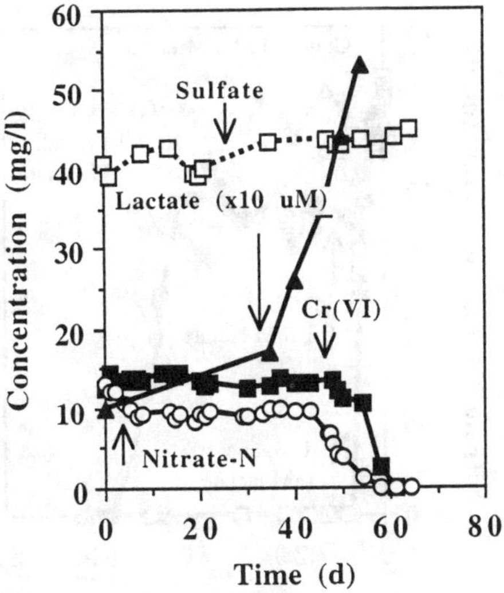

FIGURE 2. Cr(VI) reduction responding to respiked lactate

experiment. Some Cr was detected as dissolved Cr(III) in the solution (Table 1). The low pH was not a desired result, because of dissolution of other metals in the system (data not shown).

TABLE 1. Percentage of Cr(VI) removal from the electron donor-added systems (incubation for 17 days)

Electron-donor-amended system	Initial [Cr(VI)] (mg/l)	Final [Cr(VI)] (mg/l)	Cr(VI) Removal (%)	Final [Cr(III)] (mg/l)
Sucrose	9.83	0	100	1.11
Lactate	7.93	0	100	0
Yeast extract	10.4	0	100	0.27
Acetate	11.1	7.27	34.5	0.02
Benzoate	11.1	9.58	13.7	0.11

Effects of lactate concentrations on anaerobic processes. In the experiment, four different lactate concentrations were evaluated for their effectiveness. The four concentrations were, 0.1 mM, 1 mM, 3 mM and 10 mM, respectively. The experiment results in Figure 1 show that Cr(VI) was completely reduced when initial lactate concentrations were above 1 mM. In the system with 0.1 mM lactate, no Cr(VI) reduction was observed.

In order to confirm effects of lactate concentrations on Cr(VI) reduction, additional lactate (0.07 - 0.1 mM each time in five successive spikes) was amended to the 0.1 mM lactate-amended system. From Figure 2 we can see that nitrate and Cr(VI) reduction was depended on the introduction of additional lactate. Both Cr(VI) (12.5 mg/l) and nitrate (13.2 mg/l) were completely reduced in the

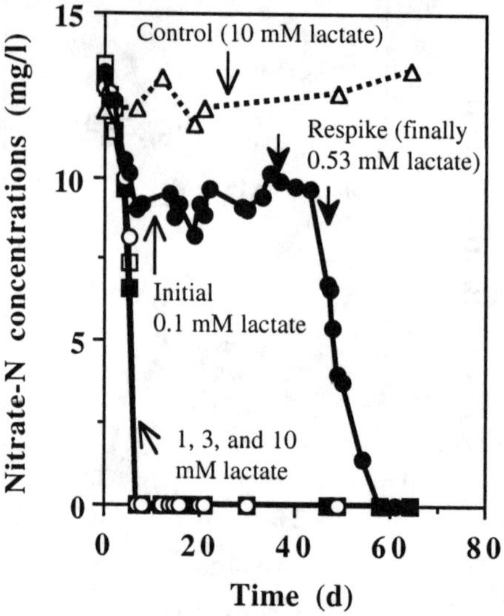

FIGURE 3. Effects of lactate concentrations on nitrate reduction

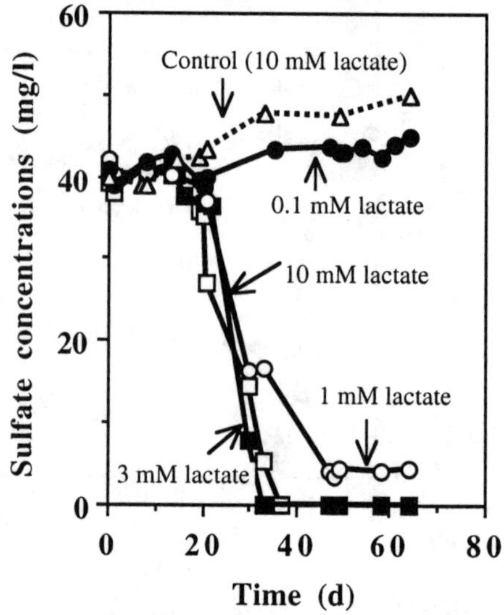

FIGURE 4. Effects of lactate concentrations on sulfate reduction

systems, when 0.53 mM total lactate was added to the system. No sulfate reduction was observed in this system (finally 0.53 mM lactate). Lactate concentration was a critical factor for Cr(VI) reduction in this experiment.

Nitrate reduction was observed in the systems with 1, 3, 10 mM initial lactate concentrations (Figure 3). The observed results were similar among these systems, with nitrate completely reduced after one week. In the 0.1 mM lactate-amended system, after nitrate was reduced to about 9 mg/l from 13 mg/l, concentrations of nitrate were stable (Figure 3). Only after additional lactate was amended to the system, did nitrate reduction re-occur (Figures 2 and 3). It suggests that lactate availability was a critical factor for the reduction of nitrate and Cr(VI) in the test systems.

The experimental results also showed that lactate concentration was critical for sulfate reduction in the system. Sulfate in the systems with 3 and 10 mM lactate was totally reduced by a 5-week period (Figure 4). Sulfate in the 1 mM lactate-amended system was reduced from 40 mg/l to about 5 mg/l and then, due to exhaustion of the lactate, was stable. In order to confirm this, the systems were analyzed for organic acids. It was found that no lactate or its products, acetic and propionic acids, were detected in the 1 mM lactate-amended system, while the 3 mM and 10 mM lactate-amended systems still had products of lactate metabolism (acetic and propionic acids). Sulfate reduction was not observed in the 0.1 mM lactate-amended system (Figure 4). No lactate or other organic acids were detected in this system.

Nitrate, Cr(VI), and sulfate are observed to be sequentially reduced in the experimental systems with sufficient concentrations of lactate. This can be seen from Figure 5. The results suggest that nitrate was reduced, before the microbial Cr(VI) reduction occurred.

Kinetics of Cr(VI) reduction. The results shown in Figure 1, indicate that Cr(VI) reduction did not start at the beginning of the experiments, but at some later

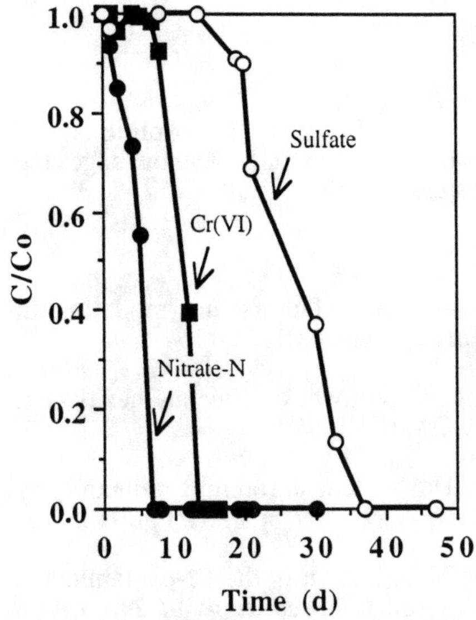

FIGURE 5. Reduction sequence of nitrate, Cr(VI), and sulfate. C_o and C are initial and current concentrations

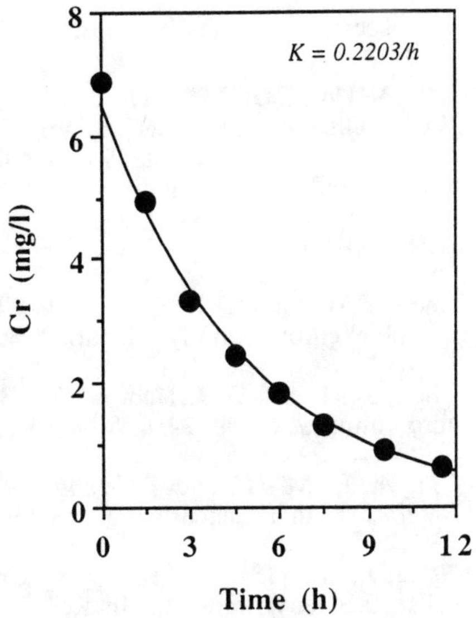

FIGURE 6. Cr(VI) reduction after the respike

stage in which the system developed an appropriate anaerobic environment. Once activity was initiated, Cr(VI) reduction was relatively fast. In order to investigate the kinetics of the Cr(VI) reduction processes, an additional experiment was conducted after the initial Cr(VI) had been totally reduced. The observed reduction of the respiked Cr(VI) appeared to match a first order kinetic process (Figure 6). The reaction kinetics can be expressed as equation 1.

$$K = \frac{1}{t}\ln\frac{C_o}{C} \tag{1}$$

where K = reduction rate (h^{-1})
t = reaction time (h)
C_o = initial Cr(VI) concentration (mg/l)
C = current (at time t) Cr(VI) concentration (mg/l).

First order rate constant for the removal of the respiked Cr(VI) was estimated to be 0.2203 h^{-1}.

CONCLUSIONS

Addition of sufficient electron donors in the experimental systems can effectively enhance Cr(VI) reduction. Approximate 10 mg/l of Cr(VI) was completely reduced in the sucrose-amended, the yeast extract-amended, and the lactate-amended systems during the 17-day experiments.

Insufficiency of available electron donors in subsurface environments is a critical factor for biotreatment of Cr(VI). It may be necessary to input additional electron donors for the remediation of Cr(VI)-contaminated sites.

The utilization sequence of nitrate, Cr(VI) and sulfate as electron acceptors, was observed in the test systems.

ACKNOWLEDGMENTS

The research was supported by the U.S. EPA. The opinions and conclusions do not necessarily reflect those of the agency. J. H. acknowledges the associate fellowship from the National Research Council.

REFERENCES

Chen, J.M., and O.J. Hao. 1996. "Environmental factors and modeling in microbial chromium(VI) reduction." *Wat Environ. Res.* 68:1156-1164.

Chen, J. M. and O. J. Hao. 1997. "Biological removal of aqueous hexavalent chromium." *J. Chem. Tech. Biotechnol.*, 69:70-76.

Chirwa, E. M. N., and T. Wang. 1997. "Hexavalent chromium reduction by *Bacillus* sp. In a packed-bed bioreactor." *Environ, Sci. Technol.,* 31:1446-1451.

Hong, J., and G.W. Sewell. 1998. "Cr(VI) bioreduction in the Cr-contaminated subsurface environment." In R.P. Paddock (ed.), *Proceedings of 216th ACS National Meeting - Environmental Chemistry Division.* 38(2): 7-9.

Krauter, P., R. Martinelli, K. Williams, and S. Martins. 1996. "Removal of Cr(VI) from ground water by *Saccharomyces cerevisiae.*" *Biodegrad.*, 7:277-286.

Lovley, D. R., and E.J.P. Phillips. 1994. "Reduction of chromate by *Desulfovibrio vulgaris* and its c3 cytochrome." *Appl. Environ. Microbiol.*, 60:726-728.

Shen, H., P.H. Pritchard, and G.W. Sewell. 1996. "Microbial reduction of Cr(VI) during anaerobic degradation of benzoate." *Environ. Sci. Technol.*, 30:1667-1674.

Wang, P., M. Tsukasa, K. Komori, M. Sasatsu, K. Toda, and H. Ohtake. 1989. "Isolation and characterization of an *Enterobacter cloacae* strain that reduces hexavalent chromium under anaerobic conditions." *Appl. Environ. Microbiol.*, 55:1665-1669.

Wang, Y., and C. Xiao. 1995. "Factors affecting hexavalent chromium reduction in pure cultures of bacteria." *Wat. Res.*, 29:2467-2474.

IN SITU MICROBIAL VOLATILIZATION OF SELENIUM IN SOILS: A CASE HISTORY

Peter T. Zawislanski and Sally M. Benson (LBNL, Berkeley, California)
Gamani R. Jayaweera and L. Wu (University of California, Davis, California)
William T. Frankenberger (University of California, Riverside, California)

ABSTRACT: A pilot-scale field experiment has been conducted since 1990 to test the effectiveness of microbial volatilization in removing selenium (Se) from soils contaminated with agricultural drainage water. The experiment, in which only irrigation and aeration were employed to enhance microbial processes, was designed to measure all major Se fluxes, including not only selenium loss via volatilization, but also advection with infiltrating rainwater, evapotranspirative transport, and plant uptake. The goal was to account for the total Se mass balance and address questions as to the significance of microbial volatilization relative to other fluxes. Although data collected from 1990 to 1994 showed decreases of Se concentrations in the top soil, subsequent data demonstrated that advective Se fluxes due to rainwater infiltration and evapotranspiration are largely responsible for the observed changes. Se volatilization was measured to account for an annual loss of only about 1%, with volatilization rates decreasing significantly with time, presumably due to the depletion of soil organic carbon. The integrated results of this project demonstrate the advantages and even necessity of an inter-disciplinary and multi-phase approach to evaluating the effectiveness of bioremediation strategies. Extreme caution needs to be taken in interpreting early results; long-term data collection and follow-up are indispensable.

INTRODUCTION

The delivery of high concentrations of selenium (Se) in agricultural wastewater to Kesterson Reservoir, CA, during the late 1970's and early 1980's resulted in one of the most prominent cases of ecosystem-scale trace element contamination of soil, water, and biota (Benson et al., 1990). Small-scale field and laboratory experiments carried out from 1987 to 1990 indicated that microbial volatilization may be an effective means of dissipating selenium from soils (Karlson and Frankenberger, 1990). However, measured gaseous Se emissions could not account for observed Se depletion in the soil, suggesting that only part of the Se decline could be due to volatilization. Discrepancies could not be resolved because the early experiments were not designed to account for solute displacement. Furthermore, small-scale experiments generally involved intensive management, such as frequent irrigation, temperature control, and organic supplements, all designed to increase microbial activity. Such management is not feasible on the scale of the 500 acre Kesterson Reservoir. Therefore, a pilot-scale field study was carried out to test the effectiveness of low-intensity management strategies, including combinations of irrigation, disking, and vegetation. A synthesis of data collected over an eight-year-long experiment is analyzed with the intent of evaluating the effectiveness of low-cost soil management as a means of Se bioremediation.

MATERIALS AND METHODS

Site Layout. A 2-acre site was selected in an area which was very frequently flooded during the operation of the Reservoir and supported primarily cattail (*Typha*) vegetation. Se concentrations in this area (up to 100 μg g^{-1}) were amongst the highest in the Reservoir. Cattail residues on the soil surface were incorporated into the top 20 cm of soil by repeated disking and rototilling, in order to homogenize the surface soils and improve the availability of cattail litter as a carbon source. The plot was then divided into four subplots, each being reserved for a particular treatment: irrigation only ("I"), irrigation and disking ("ID"), disking only ("D"), and control or no action ("C") (Fig. 1). The irrigated plots were watered once a week with 10 mm of water. The disked plots were rototilled every 2 weeks, except during the rainy season. After July 1996, the treatments were terminated.

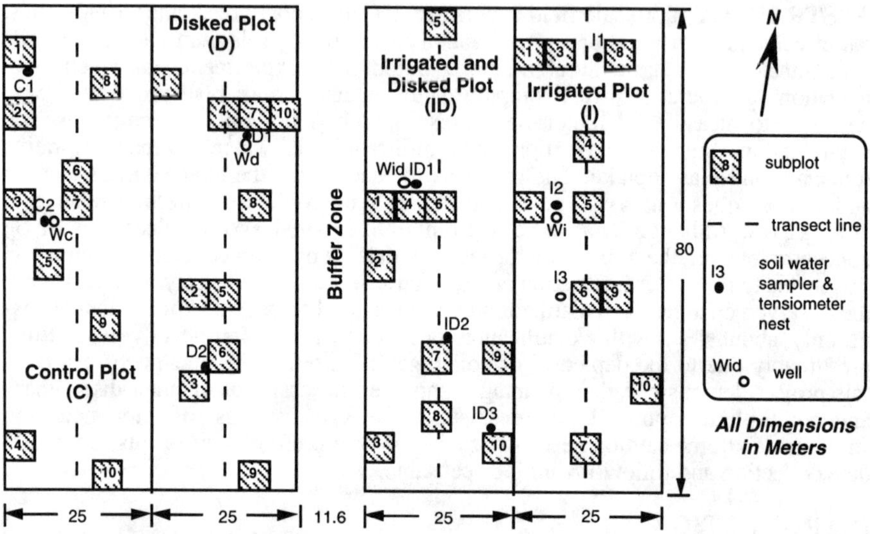

FIGURE 1. Layout of pilot-scale experimental site.

Soil Sampling and Analysis. An annual north-south transect was taken through the middle of each treatment in July from 1990 to 1996. Follow-up sampling took place in treatments I and ID in August 1998. Samples were obtained down to the depth of 0.60 m, in 0.15 m increments. Soil samples were extracted for total Se using an acid digest (Tokunaga et al., 1994) and analyzed via hydride-generation atomic absorption spectroscopy HGAAS (Weres et al., 1989). Soils were also extracted using a 1:5 soil:water extract and analyzed for total soluble Se via HGAAS (Tokunaga et al., 1994).

Direct Measurements of Volatile Se Emissions. Seven 5 m by 5 m subplots were randomly selected from each treatment to measure in-situ volatile Se emissions, using volatilization chambers (Biggar and Jayaweera, 1993). Measurements were made on a monthly basis during the spring, summer, and fall. Rates were not measured in the winter due to inclement weather conditions, but past experience showed very low volatilization rates during that time. Measurements were taken over an area of 0.25 m^2 for 24 hours, thereby accounting for diurnal temperature fluctuations. Volatile Se collected on activated carbon was extracted and analyzed using HGAAS.

Uptake of Se by Plants. Measurements of biomass and plant tissue-Se (Wu et al., 1993) were performed on plants at the field site on an annual basis.

Moisture and Solute Monitoring. Ten vadose zone monitoring nests, each consisting of four tensiometers and four soil water samplers, were installed down to a depth of 1 m (Fig. 1). All of the instruments were buried at least 30 cm below the soil surface in order to allow disking machinery to pass over the soil. Soil water samples and tensiometer readings were taken on a monthly basis from 1990 through 1994 and quarterly through 1996. Soil water was analyzed for total dissolved Se using HGAAS. Four wells were sampled on a quarterly basis. Neutron probe readings were taken in these well casings at monthly intervals.

RESULTS AND DISCUSSION

Soil-Se and Salinity Levels. Selenium concentrations in the top 0.60 m of soil in the irrigated treatment are shown in Fig. 2. Based on these data, and similar trends from the irrigated/disked and disked plots, it appeared that there were significant decreases in Se concentrations in the top 0.15 m interval, but slight increases in Se in the deeper intervals of the treated plots from 1990 to 1994. From 1994 to 1996, Se concentrations in the top 0.15 m of all plots, including the control treatment, increased. When differences in bulk density of the soils are considered and the mass of Se in the top 0.60 m is integrated (Fig. 3), the early decreases in Se appear much less significant. Dates listed in each column represent the years from which the given year was statistically different (95% confidence level). This shows that following the initial decline from 1990 to 1991, very few significant differences were observed. The follow-up sampling in 1998 showed a decrease in Se and salinity (Fig. 4; expressed as electrical conductivity of soil:water extracts) in the irrigated treatment, but no change in the irrigated/disked treatment. The overall similarity in inter-treatment Se trends, as well as matching changes in salinity, suggest processes unrelated to microbial volatilization are responsible for Se fluxes.

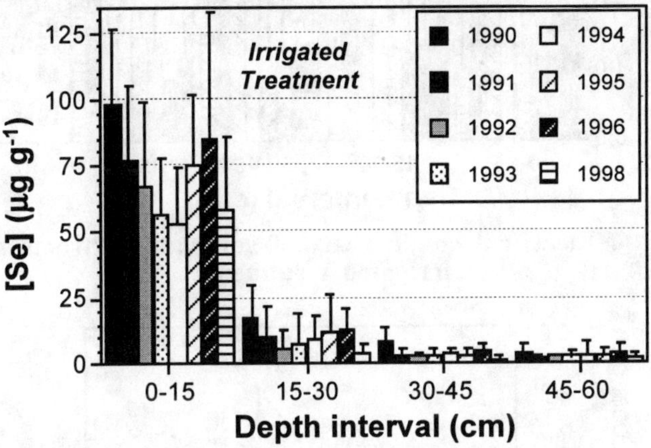

FIGURE 2. Concentrations of Se in the irrigated treatment.

Volatile Se Emissions. Cumulative Se volatilization, as measured during 1993, is shown in Fig. 5. Se losses, expressed as mass per area, can be compared with data in Fig. 3. Annual Se emissions, measured using this method, are two to three orders of magnitude lower than the total Se inventory, and therefore cannot account for the decreases in Se in the top 0.15 m as shown in Fig. 1. Although seasonal and spatial variability was observed, in none of the treatments did volatile emissions amount to more than 1% of the Se inventory in the 0-60 cm interval.

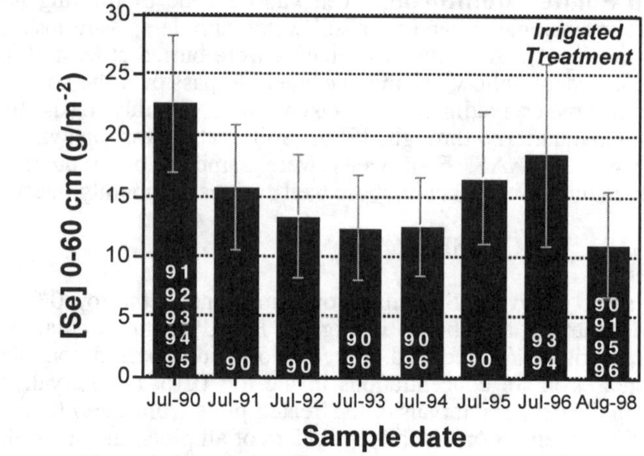

FIGURE 3. Cumulative total Se mass, normalized to area, in the 0-60 cm depth interval in the irrigated treatment.

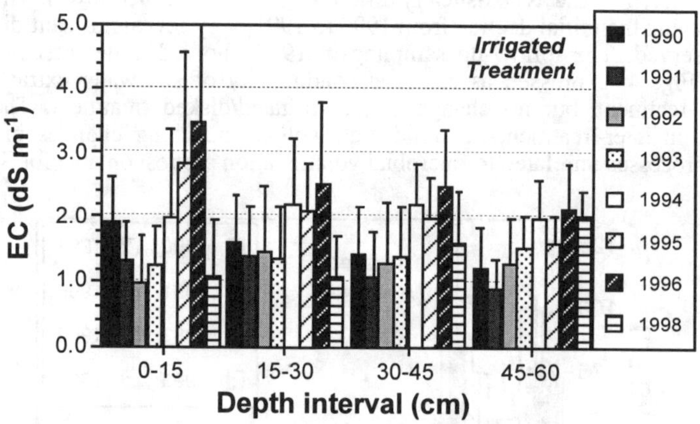

FIGURE 4. Electrical conductivity of soil:water extracts from the irrigated treatment.

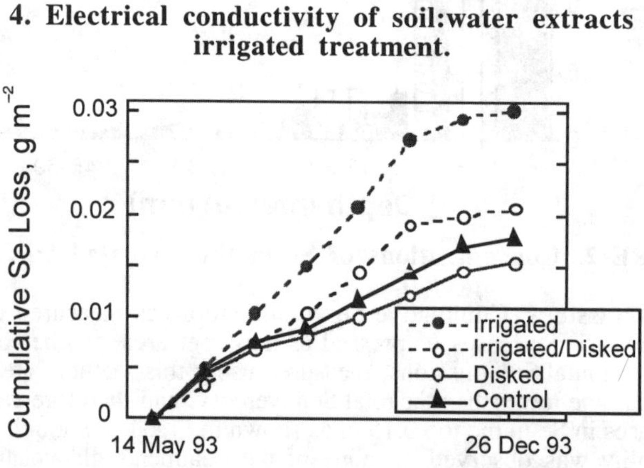

FIGURE 5. Cumulative volatile Se emissions in 1993.

Uptake of Se by Plants. The presence of plants has a profound effect on the soil-water regime by redistributing solutes along rhizosphere-directed hydraulic potential gradients. In addition, plants take up solutes in soil water, including Se. Plants in treatment I accumulated the highest Se mass (3.22 mg Se m^{-2} yr^{-1}). A comparison of the Se flux into plant tissue (1 to 3 mg Se m^{-2} yr^{-1}) with the total Se inventory in the soil (Fig. 3), indicates that plants remove less than 0.05% of the total Se per year.

Transport of Se in the Aqueous Phase. All treatments were affected by winter and spring rainfall, which in the case of the vegetated treatments resulted in as much as a tripling of moisture content. The vast and rapid changes in moisture content following major rainfall events strongly influence solute transport into the soil profile. Changes can occur over relatively short periods of time. For example, during one storm in March 1992, rainwater infiltrated *as deep as 1.6 m*. Such deep infiltration results in the displacement of Se-rich pore-waters, as seen in Fig. 6, where as much as 2 g of Se per m^2 has been displaced in the aqueous phase.

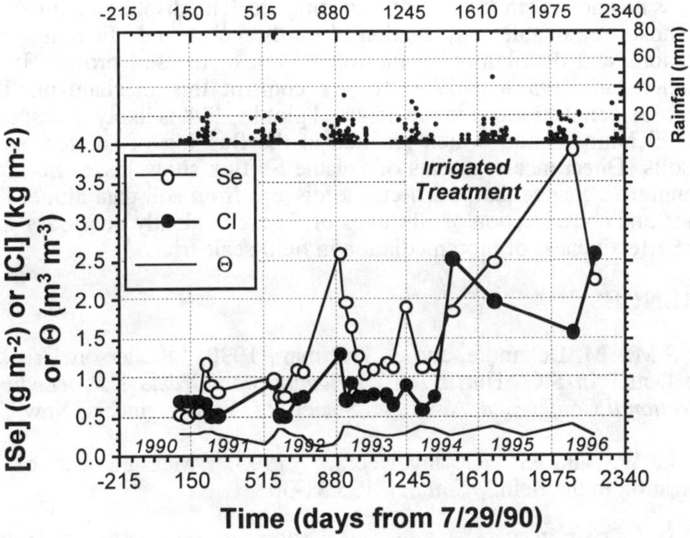

FIGURE 6. Mass of Se and Cl, and moisture content (Θ) over the depth of 0.4 m to 1.0 m in treatment I.

Data Synthesis. The mean mass contributions of the processes involved are summarized in Table 1 for the irrigated treatment. The goal of presenting data in this fashion is to elucidate whether or not the measured and estimated fluxes can account for the observed changes in total Se inventory in the 0-60 cm soil interval. The second to last column shows the amount of Se which is not accounted for after volatilization and plant accumulation are considered and which would need to be leached out of the 0-60 cm interval in order to reconcile the mass balance. The last column shows the amount of soluble Se potentially available for leaching by infiltrating rainwater. This analysis is conservative in that it does not consider the possibility of in-situ Se solubilization (Tokunaga et al., 1994). From this presentation it is clear that the mass of soluble Se is the same order of magnitude as the mass required to complete the Se mass balance. It is also apparent that Se volatilization and plant uptake are very small fluxes compared to vertical fluxes of aqueous Se.

Table 1. Annual losses of Se fractions in the 0-60 cm soil interval in treatment I and estimates of soluble Se available for leaching.

Period	Δ in total Se	Se volatilized		Se accumulated in plants		Se required to be flushed		Soluble Se
		low est	high est	low est.	high est.	low est.	high est.	
	(g/m2)	(g/m2)	(g/m2)	(g/m2)	(g/m2)	(g/m2)	(g/m2)	(g/m2)
1990-91	-6.9	0.09	0.27	0.005	0.05	*6.81*	*6.58*	***2.06***
1991-92	-2.4	0.09	0.135	0.005	0.05	*2.31*	*2.22*	***1.56***
1992-93	-1	0.034	0.051	0.005	0.05	*0.96*	*0.90*	***2.30***
1993-94	0.2	0.03	0.03	0.005	0.05	*-0.24*	*-0.28*	***1.72***

CONCLUSION

It is apparent that by accounting only for volatile Se emissions and plant uptake, the Se mass balance is, in most cases, incomplete. The displacement of soluble Se is the most reasonable explanation for the observed fluctuations in Se concentrations and distribution in the top 60 cm of the soil profile. Increases in concentrations in 1995 and 1996 further confirm this mechanism. The sharp decrease in Se concentrations in the irrigated plot in 1998 is likely in response to the very wet El Niño winter, which resulted in the flushing of solutes out of near-surface soils. Direct measurements of volatile Se flux show losses no greater than 1% per annum, a change that is difficult to discern from soil data alone. Long-term monitoring and consideration of all fluxes of Se are evidently necessary in order to assess the effectiveness of bioremediation in field-scale trials.

REFERENCES

Benson, S.M., M. Delamore, and S. Hoffman. 1990. "Kesterson Crisis: Sorting out the facts." In S.C. Harris (ed.) *Irrigation and Drainage Proceedings of the 1990 National Conference*. American Society of Civil Engineers. New York.

Biggar, J. W. and G. R. Jayaweera. 1993. "Measurement of selenium volatilization in the field." *Soil Sci.* 155:31-36.

Karlson, U. and W. T. Frankenberger, Jr. 1990. "Volatilization of selenium from agricultural evaporation pond sediments." *Sci. Total Environ.* 92:41-54.

Weres, O., A.-R. Jaouni, and L. Tsao. 1989. "The distribution, speciation and geochemical cycling of selenium in a sedimentary environment, Kesterson Reservoir, California, U.S.A." *Appl. Geochem.* 4:543-563.

Wu, L.; Enberg, A. W.; Tanji, K. K. 1993. "Natural establishment and selenium accumulation of herbaceous plant species in soils with elevated concentrations of selenium and salinity under irrigation and tillage practice." *Ecotox. and Environ. Safety* 25, 127-140.

Tokunaga, T.K, P.T. Zawislanski, P.W. Johannis, S. Benson, and D.S. Lipton. 1994. "Field investigations of selenium speciation, transformation, and transport in soils from Kesterson Reservoir and Lahontan Valley." In, *Selenium in the Environment*. W.T. Frankenberger and S. Benson, eds. Marcel Dekker. New York. pp. 119-138.

BIOLOGICAL TREATMENT OF PERCHLORATE CONTAMINATED WATERS

Bruce E. Logan, Kijung Kim, Joel Miller, Peter Mulvaney, and Richard Unz
(The Pennsylvania State University, University Park, PA)

ABSTRACT: The discovery of perchlorate contamination of drinking water sources has created a national crisis due to perchlorate's stability and persistence in conventional drinking water treatment systems. It is demonstrated here that perchlorate can be degraded by microorganisms under anoxic conditions using perchlorate as an electron acceptor, and either acetate or hydrogen gas as electron donors. Perchlorate was removed to below detectable levels (<4 µg/L) in fixed bed reactors packed with activated carbon as a support medium when acetate was fed in excess to the reactors. When the reactor was temporarily mixed to redistribute the biomass in the column, acetate was completely removed, resulting in perchlorate release likely due to desorption from the carbon. Preliminary suspended growth reactor experiments demonstrated that hydrogen gas could also support perchlorate-degrading cultures.

INTRODUCTION

Ammonium perchlorate is a highly oxidized compound that has been produced in large quantities for use in solid rocket propellent. Perchlorate has already been measured in several surface and ground waters in many states including: Arizona, Arkansas, California, Indiana, Iowa, Louisiana, Maryland, Nevada, New York, Pennsylvania, New York, Texas, Utah, Virginia, and West Virginia. The EPA expects perchlorate to be found in virtually all of the lower 48 states except perhaps Maine, Vermont, Connecticut, and Rhode Island, and that it may affect the water supplies of over 12 million people. Perchlorate is a health concern due to its interference with iodine in the production of hormones in the thyroid. An interim drinking water action level has been set by California at 18 µg/L.

Perchlorate can serve as an electron acceptor for a number of different bacterial strains, and therefore it can be microbially degraded under anoxic conditions (Logan, 1998a). In order to remove perchlorate from drinking water, we have proposed two different biological reactors for drinking water treatment: a fixed bed reactor with the capability of biofilm regeneration and intermittent bed mixing; and an unsaturated multiphase biological reactor supported by a hydrogen oxidizing bacterial consortium (Logan, 1998b). Preliminary results are presented here demonstrating that perchlorate can be degraded to below detectable levels in fixed bed acetate-fed reactors, but that acetate must be used in excess. Hydrogen gas has been considered for use in denitrifying bioreactors for water treatment, and is an attractive alternative to acetate due to its low solubility in water. We therefore conducted tests to demonstrate that hydrogen gas can serve as an electron donor for perchlorate reduction.

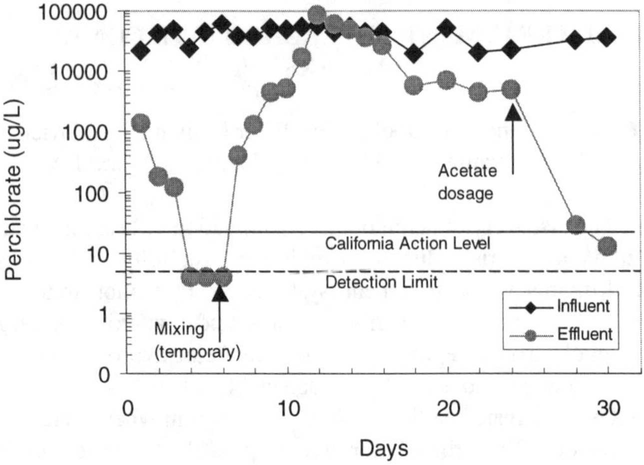

Figure 1. Perchlorate reduction in a fixed-bed acetate-fed reactor. Mixing (fluidization) was only conducted over a few minutes to mix the reactor contents.

MATERIALS AND METHODS

A bench-scale fixed bed reactor (244 cm high by 7.6 cm diameter) constructed from acrylic tubing was filled to 96 cm with activated carbon (Carbon 1240, Norit, OK). The reactor was inoculated with a perchlorate-acclimated culture, suspended in acetate and trace-mineral medium (van Ginkel et al., 1995), developed from primary digester sludge obtained from a local wastewater treatment plant. After one day, the column was operated in upflow mode and continuously fed the AGW and media at a hydraulic loading rate of 0.11 gpm/ft^2. AGW consisted of tapwater (groundwater source) aerated to removed chlorine, and nutrients (nitrogen and phosphorus). Media was made up in concentrated (100×) and diluted to final concentrations of ~35 mg/L of perchlorate and either 30 or 60 mg/L of acetate. Concentrations of perchlorate and acetate were measured using either a Dionex 100 or 500 Ion Chromatograph using an AS-11 column. For perchlorate measurements, we followed procedures in Dionex Application Note 121.

A hydrogen-oxidizing, perchlorate reducing culture was developed in 125 mL crimp-sealed serum bottles inoculated with activated sludge from a local wastewater treatment plant. The medium consisted of the following constituents 300 mg/L $NaClO_4^-$; 1550 mg/L K_2HPO_4; 850 mg/L NaH_2PO_4; 500 mg/L $NH_4H_2PO_4$; 3000 mg/L Na_2CO_3; and a trace mineral solution. Following inoculation of the bottles, headspace volumes were purged with 5% H_2/95% N_2 gas. After acclimation, evidenced by perchlorate reduction, the contents of a bottle were transferred into a sterile reactor (Applicon, Inc.) and fed a mixture of hydrogen and carbon dioxide gases. After 12 days, the reactor was operated at a detention times of 16.7 hr.

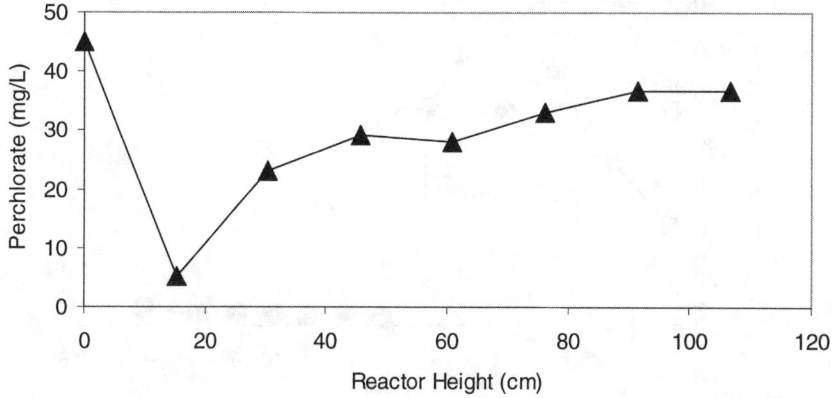

Figure 2. Perchlorate concentration profile within the fixed-bed reactor on day 15.

RESULTS AND DISCUSSION

Fixed bed Reactor. Within just four days of operation, perchlorate was removed to below detectable limits (<4 µg/L) in the activated carbon packed bed reactor (Figure 1). Other results in our laboratory using smaller and shorter columns (Logan and Kim, 1998) indicated that this level of performance of the reactor could be maintained for weeks but that eventually the column would begin to clog due to the accumulation of biomass (data not shown). Therefore, we wanted to examine what the effect of backwashing would be on the performance of the column. We mixed (fluidized) the column on day 6 using tap water, let the media settle, and then re-initiated column operation.

After media mixing, the column performance deteriorated, eventually resulting in no perchlorate removal by day 12. With continued operation, perchlorate levels again began to drop but did not do so within four days as we originally observed. Intracolumn testing on day 15 revealed that perchlorate was being reduced to 5 mg/L within a short distance in the column (12 cm), but that perchlorate concentrations actually increased at subsequent sampling locations (Figure 2). The increase in perchlorate concentrations suggested that perchlorate was desorbing from the column and that biological activity was insufficient to remove the perchlorate. Thus, perchlorate was desorbing at a rate faster than it could be used by the bacteria.

By sampling the effluent for acetate, we realized that acetate was no longer present in excess, and on day 24 we doubled the concentration of acetate in the feed. Within a few days perchlorate concentrations were reduced to almost 10 µg/L, or to below the 18 µg/L interim State of California drinking water standard (Figure 1).

Hydrogen-oxidizing Reactor. Perchlorate reduction was found to be accomplished under anoxic and hydrogen oxidizing conditions. After about 7 days of operation, the

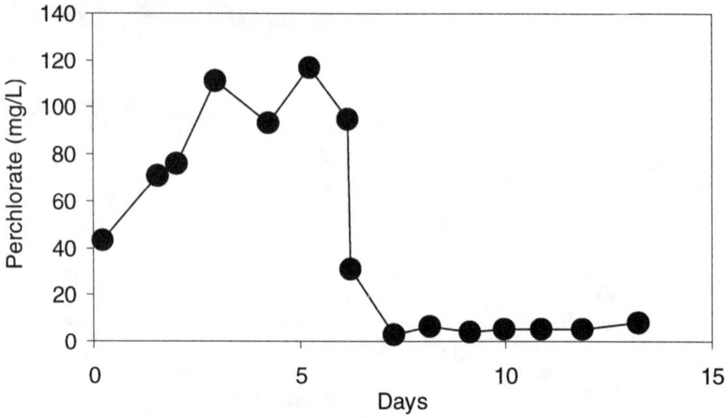

Figure 3. Perchlorate reduction by a hydrogen-oxidizing bacterial consortium in a chemostat.

suspended-growth reactor operated in chemostat mode reduced perchlorate by 240 mg/L, resulting in effluent perchlorate concentrations of 3 to 8 mg/L (Figure 3). Although these treatment levels would not be acceptable for drinking water treatment, we demonstrated that it was possible to achieve perchlorate reduction under hydrogen oxidizing conditions by perchlorate-acclimated cultures. It has similarly been shown that perchlorate could be degraded under hydrogen oxidizing conditions using chlorate-degrading mixed cultures (van Ginkel et al., 1995), indicating that hydrogen-fed reactors may have potential for drinking water treatment. Such reactors have only been previously investigated for nitrate removal (Rittmann, 1995).

CONCLUSIONS

It has been demonstrated that perchlorate can be removed to below detectable concentrations in acetate-fed fixed bed reactors. As a part of ongoing work, other electron donors, such as methanol, ethanol and lactate will be examined. The oxidation of hydrogen can be used to support microbial reduction of perchlorate, but further work with fixed-bed reactors will be necessary to achieve perchlorate reductions sufficient to meet anticipated drinking water standards.

ACKNOWLEDGMENTS

This research has been supported in part by the National Science Foundation (BES-9714575), the American Water Works Association Research Foundation (AWWARF, No. 2557), and The Pennsylvania State University.

REFERENCES

Logan, B.E. 1998. "Process for treating perchlorate-contaminated drinking water." Patent application, The Pennsylvania State University.

Logan, B.E. 1998. "A review of chlorate and perchlorate respiring microorganisms." *Bioremediation J.* 2(2):69-79.

Logan, B.E., and K. Kim. 1998. "Microbiological treatment of perchlorate contaminated groundwater." *Proceedings of the Southwest focused groundwater conference: Discussing the issue of MTBE and perchlorate in the ground water*, Anaheim, CA, June 3-4. National Ground Water Association, pp. 87-90.

Rittmann, B.E. 1995. "Fundamentals and application of biofilm processes in drinking water treatment." *Handbook of Environmental Chemistry*, Vol 5B, J. Hrubec, ed., pp. 61-87.

van Ginkel, G.C., A.G.M. Kroon, G.B. Rikken, and S.W.M. Kengen. 1998. "Microbial conversion of perchlorate, chlorate and chlorite." *Proceedings of the Southwest focused groundwater conference: Discussing the issue of MTBE and perchlorate in the ground water*, Anaheim, CA, June 3-4. National Ground Water Association, pp. 92-95.

van Ginkel, G.C., C.M. Plugge, and C.A. Stroo. 1995. "Reduction of chlorate with various energy substrates and innocula under anaerobic conditions." *Chemosphere 31*: 4057-4066.

AUTHOR INDEX

This index contains names, affiliations, and volume/page citations for all authors who contributed to the eight-volume proceedings of the Fifth International In Situ and On-Site Bioremediation Symposium (San Diego, California, April 19–22, 1999). Ordering information is provided on the back cover of this book.

The citations reference the eight volumes as follows:

5(1): Alleman, B.C., and A. Leeson (Eds.), *Natural Attenuation of Chlorinated Solvents, Petroleum Hydrocarbons, and Other Organic Compounds*. Battelle Press, Columbus, OH, 1999. 402 pp.

5(2): Leeson, A., and B.C. Alleman (Eds.), *Engineered Approaches for In Situ Bioremediation of Chlorinated Solvent Contamination*. Battelle Press, Columbus, OH, 1999. 336 pp.

5(3): Alleman, B.C., and A. Leeson (Eds.), *In Situ Bioremediation of Petroleum Hydrocarbon and Other Organic Compounds*. Battelle Press, Columbus, OH, 1999. 588 pp.

5(4): Leeson, A., and B.C. Alleman (Eds.), *Bioremediation of Metals and Inorganic Compounds*. Battelle Press, Columbus, OH, 1999. 190 pp.

5(5): Alleman, B.C., and A. Leeson (Eds.), *Bioreactor and Ex Situ Biological Treatment Technologies*. Battelle Press, Columbus, OH, 1999. 256 pp.

5(6): Leeson, A., and B.C. Alleman (Eds.), *Phytoremediation and Innovative Strategies for Specialized Remedial Applications*. Battelle Press, Columbus, OH, 1999. 340 pp.

5(7): Alleman, B.C., and A. Leeson (Eds.), *Bioremediation of Nitroaromatic and Haloaromatic Compounds*. Battelle Press, Columbus, OH, 1999. 302 pp.

5(8): Leeson, A., and B.C. Alleman (Eds.), *Bioremediation Technologies for Polycyclic Aromatic Hydrocarbon Compounds*. Battelle Press, Columbus, OH, 1999. 358 pp.

Aagaard, Per (University of Oslo/NORWAY) 5(1):213

Abbondanzi, F. (Centro Ricerche Ambientali Montecatini/ITALY) 5(3):71

Abbott, James E. (Battelle/USA) 5(3):201; 5(5):57

Abdelouas, Abdesselam (University of New Mexico/USA) 5(4):59, 115

Abrajano, Teofilo (Memorial University of Newfoundland/CANADA) 5(1):207

Abriola, Linda M. (University of Michigan/USA) 5(1):201

Acomb, Lawrence (Geosphere, Inc./USA) 5(3):505

Adamek, Scott (AGI Technologies/USA) 5(3):189

Adnanes, Jenny (University of Oslo/NORWAY) 5(1):213

Adrian, Neal (U.S. Army/USA) 5(7):1

Agar, John G. (O'Connor Associates Environmental Inc./CANADA) 5(5):111

Aitchison, Eric W. (Ecolotree, Inc./USA) 5(6):139

Al-Abed, Souhail R. (U.S. EPA/USA) 5(2):263
Al-Awadhi, Nader (Kuwait Institute for Scientific Research/KUWAIT) 5(5):13
Al-Daher, Reyad (Kuwait Institute for Scientific Research/KUWAIT) 5(5):13
Al-Hakak, A. (Biotechnology Research Institute/CANADA) 5(3):439
Alleman, Bruce (Battelle/USA) 5(2):7; 5(3):201
Allen, Harry L. (U.S. EPA/USA) 5(6):89
Alsaleh, Esmaeil (King's College/UK) 5(8):111
Altman, Denis J. (Westinghouse Savannah River Company/USA) 5(2):107
Alvarez, Pedro J. (University of Iowa/USA) 5(1):171; 5(4):79; 5(6):139
Ambert, Jack (Battelle Europe/SWITZERLAND) 5(1):159
Ambus, Per (Risø National Laboratory/DENMARK) 5(6):15
Amerson, Illa L. (Oregon Graduate Institute/USA) 5(3):123
Ampleman, Guy (National Defense/CANADA) 5(7):15
Anasawa, T.A. (Campinas State University-UNICAMP/BRAZIL) 5(8):105
Anderson, Bruce N. (RMIT University/AUSTRALIA) 5(3):517
Anderson, David W. (Roy F. Weston, Inc./USA) 5(2):157
Andreotti, Giorgio (ENI/AGIP S.p.A./ITALY) 5(1):159; 5(5):69, 183
Andretta, Massimo (Centro Ricerche Ambientali Montecatini/ITALY) 5(3):511
Andrews, Elizabeth J. (University of Minnesota/USA) 5(2):35
Andrews, Scott (SECOR International Inc./USA) 5(1):71
Angelmi, Barbara (ARBES-Umwelt GmbH/GERMANY) 5(3):487
Anthony, John W. (Parsons Engineering Science, Inc./USA) 5(1):121
Antia, Jimmy (University of Cincinnati/USA) 5(8):283
Araneda, Edgardo (Jordforsk/NORWAY) 5(8):99
Arienzo, Michele (Universita degli Studi di Napoli/ITALY) 5(7):197, 203
Arlotti, Daniele (Foster Wheeler Environmental Italia S.R.L./ITALY) 5(5):183
Arrocha, Alfonso (PDVSA Manufactura y Mercadeo/VENEZUELA) 5(3):215
Ashcom, Dave W. (AGI Technologies/USA) 5(3):189
Atalay, Ferhan Sami (Ege University/TURKEY) 5(5):149
Atwater, Jim (University of British Columbia/CANADA) 5(1):231
Auria, Richard (Institut de Recherche pour le Developpement/FRANCE) 5(3):31
Autenrieth, Robin L. (Texas A&M University/USA) 5(3):209; 5(6):57; 5(7):21
Aziz, Carol E. (Groundwater Services, Inc./USA) 5(1):83

Babel, Wolfgang (Umweltforschungszentrum/GERMANY) 5(7):143, 235
Babin, J. (Golder Associates/HONG KONG) 5(8):309
Bächle, Arthur (Mannheimer Versorgungs und Verkehrsgesellschaft mbH (MVV)/GERMANY) 5(3):337
Baker, Ralph S. (ENSR Consulting & Engineering/USA) 5(3):149, 155
Balba, M. Talaat (Kuwait Institute for Scientific Research/KUWAIT) 5(5):13
Balcer, Denis (ARCADIS Geraghty & Miller, Inc./USA) 5(2):135
Bandala, Erick R. (Instituto Mexicano de Technologia del Agua/MEXICO) 5(7):137
Banwart, Steven A. (University of Sheffield/UK) 5(1):277, 307
Bär, Eckart (Environmental Expert Office/GERMANY) 5(3):363, 369, 517
Barbaro, A. (EniTecnologie/ITALY) 5(8):129

Barbonio, Miriam L. (Chevron Chemical Company/USA) 5(4):91
Barcelona, Michael J. (University of Michigan/USA) 5(1):171, 201; 5(3):345
Bare, Richard E. (Exxon Research & Engineering/USA) 5(3):227
Barker, Craig S. (IT Environmental Pty Ltd./SOUTH AUSTRALIA) 5(2):199
Barkovskii, Andrei (The University of Michigan/USA) 5(6):75
Barlaz, Morton A. (North Carolina State University/USA) 5(1):165
Barnes, Paul (Waste Management, Inc./ USA) 5(7):39, 45, 63
Barnett, J. Stephen (Exponent Environmental Group/USA) 5(7):83
Bartholomae, Philip G. (BP Amoco Company/USA) 5(1):97, 219
Bartlett, Craig L. (DuPont Company/ USA) 5(1):29
Baartmans, R. (TNO Institute of Environmental Sciences/ NETHERLANDS) 5(2):141
Basel, Michael D. (Montgomery Watson/ USA) 5(1):41
Beard, A.G. (Biotechnology Research Institute/CANADA) 5(3):439
Beaton, Yvonne (University of Aberdeen/ UK) 5(7):247
Beaudet, R. (Institut Armand-Frappier/ CANADA) 5(7):107
Beaulieu, Maude (INRS, Institut Armand-Frappier/CANADA) 5(7):89
Beaulieu, Sophie (École Polytechnique de Montréal/CANADA) 5(6):211
Bécaert, Valérie (École Polytechnique de Montréal/CANADA) 5(7):89
Beck, Frank P. (U.S. EPA/USA) 5(1):103
Beck, James W. (U.S. EPA/USA) 5(7):95
Becker, Bob (Colt Engineering Corporation/CANADA) 5(5):167
Becker, David J. (U.S. Army Corps of Engineers/USA) 5(3):155
Becker, James M. (Battelle PNNL/USA) 5(1):359
Becker, Jennifer G. (Lehigh University/ USA) 5(1):343
Becker, P. (Battelle Marine Sciences Laboratory/USA) 5(6):63
Bedient, Philip B. (Rice University/USA) 5(1):7
Bell, J.N.B. (Imperial College of Science, Technology, and Medicine/UK) 5(8):123
Belloso, Claudio (Facultad Católica de Química e Ingeniería/ARGENTINA) 5(3):331; 5(5):75, 177
Belote, J.G. (Campinas State University-UNICAMP/BRAZIL) 5(6):271
Belton, Darren (U.S. Navy/USA) 5(3):131, 383
Bender, Judith (Clark Atlanta University/ USA) 5(4):109
Bennetzen, Susan (VKI/DENMARK) 5(6):15
Benson, Sally M. (Lawrence Berkeley National Laboratory/USA) 5(4):141
Bergeron, Eric (Sodexen Group/ CANADA) 5(8):1
Bergersen, Ove (SINTEF Oslo/ NORWAY) 5(5):37; 5(6):187
Beveridge, Terry J. (University of Guelph/CANADA) 5(4):121
Beyer, Lothar (University of Kiel/GERMANY) 5(3):325, 487
Bhadra, Rajiv (Rice University/USA) 5(6):121
Bhatt, Manish (Academy of Sciences of the Czech Republic/CZECH REPUBLIC) 5(6):69
Bidgood, Jason B. (Parsons Engineering Science, Inc./USA) 5(1):121
Biehle, Alfred A. (DuPont Company/ USA) 5(1):35; 5(2):55
Bienkowski, Lisa A. (IT Group, Inc./ USA) 5(2):165
Biggar, Kevin (University of Alberta/ CANADA) 5(1):195
Bil, Jerzy (Military University of Technology/POLAND) 5(5):1
Bisaillon, Jean-Guy (Université du Quebec/CANADA) 5(8):1
Bittoni, A. (EniTecnologie/ITALY) 5(8):129
Bizzell, Cydney (Texas A&M University/ USA) 5(6):57

Blanchet, Denis (Institut Français du Pétrole/FRANCE) 5(8):117
Blechinger, Fred M. (Lucent Technologies Inc./USA) 5(3):149
Blokzijl, Ron (DHV Noord Nederland/ NETHERLANDS) 5(2):225
Blowes, David (University of Waterloo/ CANADA) 5(4):19
Boels, Dethmer (SC-DLO/ NETHERLANDS) 5(6):39
Bogan, Bill W. (EarthFax Development/ USA) 5(8):81
Bojanowski, Sebastian. (University of Applied Sciences/GERMANY) 5(4):97
Bolio, Paula A. (CH2M Hill/USA) 5(1):265
Bonaldo, Dino (Mission Research Corporation/USA) 5(3):163
Bonas, Andrew R. (EA Engineering, Science, & Technology, Inc./USA) 5(1):295
Bongers, A.M.T. (LUW/NETHERLANDS) 5(6):231
Bonner, James S. (Texas A&M University/USA) 5(3):209; 5(6):57; 5(7):21
Bonten, Luc T.C. (Wageningen Agricultural University/ NETHERLANDS) 5(8):265
Borazjani, Abdolamid (Mississippi State University/USA) 5(7):75
Borden, Robert C. (North Carolina State University/USA) 5(1):165
Borger, Arjan (TNO Institute of Environmental Sciences/ NETHERLANDS) 5(2):73, 141
Boronin, Alexander M. (Russian Academy of Sciences/RUSSIA) 5(3):209
Borresen, Marion (Norwegian Geotechnical Institute/NORWAY) 5(6):187
Bosgoed, H. (Logisticon Water Treatment/NETHERLANDS) 5(2):225
Bosma, Tom N.P. (TNO Institute of Environmental Sci/NETHERLANDS) 5(2):73, 141
Boucard, Tatiana (University of Aberdeen/UK) 5(7):247
Bourquin, Al W. (Camp Dresser & McKee Inc./USA) 5(2):107
Bouwens, Sandra (Vrije Universiteit van Amsterdam/NETHERLANDS) 5(3):395
Bowling, Leon (U.S. Marine Corps/USA) 5(3):201
Boxwell, Mimi (U.S. Army Corps of Engineers/USA) 5(7):7
Boyd, Thomas J. (U.S. Navy/USA) 5(8):223, 295
Boyle, Donald B. (Earth Tech Inc./USA) 5(1):241; 5(3):289
Brandt, Charles A. (Battelle PNNL/USA) 5(1):359
Brandt, Craig C. (Oak Ridge National Laboratory/USA) 5(4):1
Braukmeier, J. (Technical University of Hamburg-Harburg/GERMANY) 5(5):43
Brauner, J. Steven (Virginia Tech/USA) 5(8):19
Brecht, Maike (University of Kiel/ GERMANY) 5(3):325
Breedveld, Gijs D. (Norwegian Geotechnical Institute/NORWAY) 5(1):213
Brennan, Rachel A. (University of Illinois/USA) 5(2):21
Breton, Jérôme (National Research Council of Canada/CANADA) 5(7):15
Brigmon, Robin L. (Westinghouse Savannah River Co/USA) 5(2):107, 191
Briseid, Tormod (SINTEF Oslo/ NORWAY) 5(5):37; 5(6):187
Brito, F. (PDVSA Manufactura y Mercadeo/VENEZUELA) 5(3):215
Britto, Ronald (Ensafe Inc./USA) 5(8):37
Broholm, Kim (VKI/DENMARK) 5(2):81
Brown, Anthony (Komex H2O Science/ USA) 5(6):289
Brown, James L. (Roy F. Weston/REAC/ USA) 5(8):51
Brox, Gunter (TEKNO Associates/USA) 5(5):19

Bruce, Cristin L. (Arizona State University/USA) 5(3):37, 123
Brunner, Richard (Tetra Tech, Inc/USA) 5(7):57
Bryndzia, L. Taras (Shell Oil/USA) 5(2):263
Buc, Edmund A. (ARCADIS Geraghty & Miller, Inc./USA) 5(1):253
Buck, John K. (Civil & Environmental Consultants, Inc./USA) 5(6):115
Buenrostro-Zagal, J.F. (Ecatepec Institute of Tech.-TESE/MEXICO) 5(6):265
Buès, Michel A. (Laboratoire Environnement Géomecanique et Ouvrage/FRANCE) 5(3):433
Bugbee, Bruce (Utah State University/ USA) 5(6):127
Buijs, Alice (TNO Institute of Applied Geoscience/NETHERLANDS) 5(3):463
Bundy, Jake (University of Aberdeen/ UK) 5(3):351
Bureau, Jacques (École Polytechnique de Montréal/CANADA) 5(2):243; 5(8):185
Burlage, Robert (University of Tennessee/USA) 5(8):277
Butcher, Mark G. (Battelle PNWD/USA) 5(2):165
Butler, A.B. (Imperial College of Science, Technology, and Medicine/UK) 5(8):123
Buzea, Dan C. (Legette, Brashears & Graham, Inc./USA) 5(3):427

Caccavo, Frank (University of New Hampshire/USA) 5(4):13
Caffarel-Méndez, S. (Ecatepec Institute of Tech.-TESE/MEXICO) 5(6):265
Caldwell, Brian E. (DuPont Company/ USA) 5(1):35
Camesano, Terri (The Pennsylvania State University/USA) 5(2):231
Campbell, Colin (Macaulay Land Use Research Institute/UK) 5(3):351
Canfield, Mark (U.S. Air Force/USA) 5(1):241; 5(3):289
Canty, Marietta C. (MSE Technology Applications/USA) 5(4):31; 5(6):193, 205
Carey, Grant (Conestoga-Rovers & Associates/CANADA) 5(1):77, 325
Carmichael, Lisa M. (North Carolina A&T University/USA) 5(3):403
Carnevali, Monica (Univ. Bologna/ ITALY) 5(3):71
Carrig, Janie (U.S. Army Corps of Engineers/USA) 5(1):241
Carter, Gregory L. (Earth Tech/USA) 5(2):255
Carter, Sean R. (Matrix Environmental Technologies/USA) 5(3):65
Carvalho, Daniela F. (Campinas State University-UNICAMP/BRAZIL) 5(3):457
Cajthaml, Tomáš (Czech Academy of Sciences/CZECH REPUBLIC) 5(6):69
Chagas, E.P. (Campinas State University-UNICAMP/BRAZIL) 5(6):271
Chaki, Sanjib (U.S. EPA/USA) 5(5):87
Chang, Kuok Weai (Alex) (DSO National Laboratories/SINGAPORE) 5(3):257
Chang, Soon W. (Kyonggi University/ REP OF KOREA) 5(7):131
Chang, Yoon-Yang (Korea Institute of Science and Technology/REP OF KOREA) 5(6):9
Chang, Yun-Juan (University of Tennessee/USA) 5(3):345
Chaobal, Vikram (University of Alaska/ USA) 5(3):499, 505
Chard, Julie K. (Utah State University/ USA) 5(6):127
Chattopadhyay, Sandip (U.S. EPA/USA) 5(6):163
Chaudri, Amar M. (IACR- Rothamsted/ UK) 5(4):25
Chaw, Donna (Olds College/CANADA) 5(5):189
Chawla, Ramesh C. (Howard University/ USA) 5(7):179
Chee-Sanford, Joanne C. (University of Illinois at Urbana-Champaign/USA) 5(8):191
Chen, Yung-Ming (University of Michigan/USA) 5(1):201

Chenier, Martin (Sodexen Group/CANADA) 5(8):1
Chiarenzelli, Jeffrey R. (State University of New York at Oswego/USA) 5(7):167
Chino, Hiroyuki (Obayashi Corporation/JAPAN) 5(5):13
Chiu, Ying-Chih (National I-Lan Institute of Agric & Technol/TAIWAN – R.O.C.) 5(2):205
Cho, Dongheng (Inha University/REP OF KOREA) 5(6):245
Cho, Jong Soo (U.S. EPA/USA) 5(1):103, 109, 115
Christy, Ann D. (The Ohio State University/USA) 5(6):181
Clark, J.I. (C-CORE/CANADA) 5(2):249
Clement, T. Prabhakar (Battelle-PNNL/USA) 5(1):29, 83
Clemente, Andrea R. (Campinas State University-UNICAMP/BRAZIL) 5(8):105
Cline, Paul R. (Zeneca Corp/CANADA) 5(7):125
Cluxton, Phillip (Cluxton Instruments, Inc./USA) 5(2):263
Coffin, Richard B. (U.S. Navy/USA) 5(8):223
Colapietro, P. (EniTecnologie/ITALY) 5(8):129
Compeau, Geoffrey (URS Greiner Woodward Clyde/USA) 5(8):167
Connolly, Mark D. (Connolly Environmental/AUSTRALIA) 5(8):43
Consort, James (MSE Technology Applications, Inc./USA) 5(6):205
Corapcioglu, M. Yavuz (Texas A&M University/USA) 5(2):269; 5(6):9
Corseuil, Henry X. (Universidade Federal de Santa Catarina/BRAZIL) 5(1):135
Corsi, F.K. (Campinas State University-UNICAMP/BRAZIL) 5(3):457
Cosgriff, David (Champion International Corp./USA) 5(5):97
Crane, Cynthia E. (Virginia Polytechnic Institute & State Univ/USA) 5(7):191
Crawford, Ronald L. (University of Idaho/USA) 5(1):1
Crawford, Scott C. (Envirogen, Inc./USA) 5(6):1
Crosswell, Scott (Virginia Polytechnic Institute & State University/USA) 5(6):45
Crouse, Daniel G. (Roy F. Weston, Inc./REAC/USA) 5(6):89
Crowe, R. Kenneth (TRW Environmental Services, Inc./USA) 5(2):7
Croxford, Roxie (Environment Agency/UK) 5(3):239
Cunningham, Al B. (Montana State University/USA) 5(2):211; 5(4):13
Cuypers, Michiel P. (Wageningen Agricultural University/NETHERLANDS) 5(8):241

Dacyk, Peter I. (Parsons Engineering Science, Inc./USA) 5(2):1
Damera, Raveendra (General Physics Corp./USA) 5(3):115
Davis-Hoover, Wendy J. (U.S. EPA/USA) 5(2):263
de Bont, Jan A.M. (Wageningen Agricultural University/NETHERLANDS) 5(2):73
Deckwer, Wolf-Dieter (National Research Center for Biotechnology GBF/GERMANY) 5(5):155
Defibaugh, Stephen T. (Block Environmental/USA) 5(3):1
Deigaard, Lars (Scanrail Consult/DENMARK) 5(2):81; 5(3):541
DeJournett, Todd D. (Barr Engineering/USA) 5(4):79
de Kreuk, H. (BioSoil R&D/NETHERLANDS) 5(2):141
DeMarco, D. (Waterloo Hydrogeologic, Inc./USA) 5(2):281
Deming, Jody W. (University of Washington/USA) 5(8):301
Dendooven, Luc (CINVESTAV-IPN/MEXICO) 5(8):93
Deng, Lijun (University of New Mexico/USA) 5(4):59
Denham, Miles (Westinghouse Savannah River Co/USA) 5(2):191
deNoyelles, Frank (University of Kansas/USA) 5(1):283

De Palma, Fabio (Ecodeco S.p.A./ ITALY) 5(3):445

Derksen, Anja (AquaSense/ NETHERLANDS) 5(3):395

Dernbach, Lisa (CA Regional Water Quality Ctrl Bd/USA) 5(3):415

de Sain, H. (BdS, Management Consultancy/NETHERLANDS) 5(2):141

Deschênes, Louise (École Polytechnique de Montréal/CANADA) 5(2):243; 5(3):451; 5(5):63; 5(6):211; 5(7):89, 229; 5(8):185

Deshusses, Marc A. (University of California Riverside/USA) 5(3):13; 5(5):129

DeStefanis, Edward J. (Leggette, Brashears, & Graham/USA) 5(3):427

Dhankher, Om Parkash (University of Georgia/USA) 5(6):145

Diegor, E.J.M. (Memorial University of Newfoundland/CANADA) 5(1):207

Diehl, Susan V. (Mississippi State University/USA) 5(7):75

Di Gregorio, Salvatore (Universita della Calabria/ITALY) 5(3):511

Dijk, John (Wageningen Agricultural University/NETHERLANDS) 5(2):73

Di Leo, Cristina (EniTecnologie/ITALY) 5(8):129

Dinwiddie, Blake A. (Roy F. Weston, Inc./USA) 5(2):157

DiPalma, Luca (Universita degli Studi di Roma La Sapienza/ITALY) 5(6):157

Dlugonsky, J. (University of Lodz/ POLAND) 5(5):197

Dockter, Craig (ENTRIX, Inc./USA) 5(3):307

Doelman, Peter (IWACO BV/ NETHERLANDS) 5(6):231

Dooley, Maureen A. (Harding Lawson Associates/USA) 5(2):121

Dosani, Majid A. (IT Corporation/USA) 5(8):7, 31

Dosoretz, Carlos G. (Migal-Galilee Technological Center/ISRAEL) 5(6):69

Doucette, William J. (Utah State University/USA) 5(6):127

Doughty, Herb (U.S. Navy/USA) 5(3):89

Drake, Evelyn N. (Exxon Research & Engineering Company/USA) 5(6):27

Drake, Randy T. (USA) 5(3):375

Drescher, Melody (Battelle/USA) 5(3):201

Drew, Malcolm (Texas A&M University/ USA) 5(6):9

Drought, James (ARCADIS Geraghty & Miller, Inc./USA) 5(1):253

Druelle, Vincent (Gaz de France/ FRANCE) 5(8):117

Drummond, C. (The University of Michigan/USA) 5(1):201

Du, Yen-Han (National Chung Hsing University/TAIWAN – R.O.C.) 5(3):469

Duart, Paola (Battelle Geneva/ SWITZERLAND) 5(1):359

Dudal, Yves (École Polytechnique de Montréal/CANADA) 5(7):229

Duffy, Baxter (Inland Pollution Services, Inc./USA) 5(3):271, 319

Duijn, Rik (De Vries & van de Wiel/ NETHERLANDS) 5(6):39

Dumortier, R. (Biotechnology Research Institute/CANADA) 5(7):107

Dupont, R. Ryan (Utah State University/ USA) 5(3):183

Dupuy, Gary (URS Greiner Woodward Clyde/USA) 5(8):167

Durrant, Lucia Regina (Campinas State University-UNICAMP/BRAZIL) 5(3):457; 5(6):271; 5(8):105

Dzirasa, Kafui (Lancaster University/UK) 5(8):259

Eberheim, Jochen (University of Applied Sciences/GERMANY) 5(4):97

Eckert, Paul (Stadtwerke Düsseldorf AG/GERMANY) 5(3):475; 5(8):141

Edgehill, Richard U. (University of New England/USA) 5(5):203

Edwards, Stuart (Dames & Moore/ ENGLAND) 5(1):271

Eggen, Trine (Jordsforsk/NORWAY) 5(8):99

Eidsa, Gudmunn (SINTEF/NORWAY) 5(5):37

Ellis, Barry (CELTIC Technologies Ltd./UK) 5(1):65; 5(8):63
Ellis, David E. (Dupont Company/USA) 5(1):35, 59; 5(2):55
Eloskof, Abram (Foster Wheeler Environ. Corp./USA) 5(3):131, 383
Ely, Roger L. (University of Idaho/USA) 5(1):1; 5(2):147
Erbanová, Pavla (Academy of Sciences of the Czech Republic/CZECH REPUBLIC) 5(6):69
Esch, Joachim (Schunk Company/GERMANY) 5(4):97
Eshbaugh, Gerald (The Environment Company/USA) 5(3):277
Esparza, Fernando (CINVESTAV-IPN/MEXICO) 5(8):93
Evans, Patrick J. (AGI Technologies/USA) 5(1):331; 5(3):189

Fabiani, Fabio (EniTecnologie/ITALY) 5(8):129
Falconi, F.A. (Campinas State University-UNICAMP/BRAZIL) 5(8):105
Fang, C. (National University of Singapore/SINGAPORE) 5(3):389
Fang, Jiasong (University of Michigan/USA) 5(1):171
Fang, Yan (The University of Arizona/USA) 5(2):231
Farone, William (Applied Power Concepts, Inc./USA) 5(2):67, 287
Fava, Fabio (University of Bologna/ITALY) 5(3):313
Fayolle, Françoise (Institut Français du Pétrole/FRANCE) 5(3):25
Feitkenhauer, Heiko (Technical University of Hamburg-Harburg/GERMANY) 5(5):25
Fennell, Donna E. (Cornell University/USA) 5(2):27
Fenwick, Caroline (King's College/UK) 5(3):363
Ferdinandy, Marijke (RIZA-Institute for Inland Water Mgt and Wastewater Treatment/NETHERLANDS) 5(8):57
Fermor, Terry R. (Lancaster University/UK) 5(8):25
Fernandes, Marilda (Universidade Federal de Santa Catarina/BRAZIL) 5(1):135
Fernández-Linares, Luis (Instituto Mexicano del Petroleo/MEXICO) 5(8):93
Fernández-Sánchez, José Manuel (CINVESTAV-IPN/MEXICO) 5(7):173
Fetterolf, Glendon (Black and Veatch/USA) 5(6):45
Fettig, Joachim (Universität Hoxter/GERMANY) 5(8):141
Field, Jim A. (Wageningen Agricultural University/NETHERLANDS) 5(7):253; 5(8):69
Fields, Keith (Battelle/USA) 5(3):89; 5(5):57
Filliat, Nadia Junod (Battelle Europe/SWITZERLAND) 5(6):51
Fischman, Daniel S. (TOSCO Marketing Company/USA) 5(3):1
Fisher, R. Todd (Groundwater Services, Inc./USA) 5(2):185
Fitzgerald, Kristin (George Mason University/USA) 5(5):143
Fliermans, Carl B. (Westinghouse Savannah River Co/USA) 5(2):107
Fode, Kimberly A. (University of Wisconsin-Milwaukee/USA) 5(1):253
Foght, Julia (University of Alberta/CANADA) 5(1):195
Foote, Eric A. (Battelle/USA) 5(2):7
Foote, Martin W. (MSE Technology Applications/USA) 5(6):205
Forbort, Jon E. (Envirogen, Inc./USA) 5(7):7
Ford, Frank C. (Henderson Paddon Environmental/CANADA) 5(6):259
Fortin, Nathalie Y. (University of California Riverside/USA) 5(3):13; 5(5):129
Francis, M. McD. (Nova Chemicals/CANADA) 5(1):207
Frankenberger, William T. (University of California/USA) 5(3):383; 5(4):141
Frankenfeld, Katrin (Forschungszentrum fur Medizintechnik und Biotechnologie e. V/GERMANY) 5(5):161

Franz, Thomas (Beatty Franz & Associates/CANADA) 5(2):281
Fredrickson, Herb (U.S. Army Corps of Engineers/USA) 5(3):545
Freixo, M. Rosário (Instituto Nacional de Engenharia e Tecnologia Industrial/ PORTUGAL) 5(5):197
Friesen, Kent (Parsons Engineering Science, Inc/USA) 5(8):13
Furlow, Lillian J. (CH2M Hill/USA) 5(3):77
Furquim, F.E. (Campinas State University-UNICAMP/BRAZIL) 5(3):457

Gagliano, Wendy R.B. (The Ohio State University/USA) 5(6):181
Gagni, S. (Centro Ricerche Ambientali Montecatini/ITALY) 5(3):71
Gagnon, Josée (École Polytechnique de Montréal/CANADA) 5(3):177, 451; 5(7):89
Gamati, Samia (Sodexen Group/ CANADA) 5(8):1
Gan, Ying Dong (University of Tennessee/USA) 5(3):345
Gannon, David John (Zeneca Corp./CANADA) 5(2):281; 5(7):125
Gao, H. (Imperial College of Sci., Tech., and Med./UK) 5(8):123
Gao, Jianping (U.S. EPA/USA) 5(6):151
Gardner, John (Pelorus Environmental Biotech Corp./USA) 5(1):71
Garnier, Patrice (Institut de Recherche pour le Developpement/FRANCE) 5(3):31
Garrett, Craig A. (Air Force Institute of Technology/USA) 5(2):275
Garrett, Robert M. (Exxon Rsrch & Engrg Co/USA) 5(3):227, 493
Garrison, A. Wayne (U.S. EPA/USA) 5(6):151
Geiselbrecht, Allison (Natl. Marine Fisheries/NOAA/USA) 5(8):301
Georgiev, Plamen S. (University of Mining & Geology/BULGARIA) 5(4):103
Gerlach, Robin (Montana State University/USA) 5(2):211; 5(4):13
Gerrard, A. Mark (University of Teesside/UK) 5(5):161, 197
Gerritse, Jan (TNO Institute of Environmental Sciences/ NETHERLANDS) 5(1):189; 5(2):73, 141
Ghosh, Mriganka M. (University of Tennessee/USA) 5(7):155, 161
Ghosh, Upal (Carnegie Mellon University/USA) 5(8):289
Gibbs, James T. (Battelle/USA) 5(2):7
Gierke, John (Michigan Technological Univ/USA) 5(3):97
Gilbert, D. (PDVSA Manufactura y Mercadeo/VENEZUELA) 5(3):215
Gillespie, Rick D. (Battelle/USA) 5(2):7
Gillette, J. Seb (Stanford University/USA) 5(8):289
Gillevet, Patrick M. (George Mason University/USA) 5(5):143
Ginn, Jon S. (U.S. Air Force/USA) 5(2):7
Girio, F.M. (Instituto Nacional de Engenharia e Tecnologia Industrial/PORTUGAL) 5(5):197
Glascoe, Lee G. (The University of Michigan/USA) 5(1):201
Glaser, John A. (U.S. EPA/USA) 5(5):103; 5(8):7, 31, 209
Glass, David J. (D Glass Associates Inc/ USA) 5(6):95
Glover, L. Anne (University of Aberdeen/ UK) 5(3):363, 369; 5(7):247
Godfrey, John M. (Integrated Science & Technology Inc/USA) 5(5):183
Goeke, Patrick M. (BE&K/Terranext/ USA) 5(3):295
Goers, Marshall (Earth Tech/USA) 5(3):289
Gohda, Hiroshi (Towa Kagaku Co., Ltd./JAPAN) 5(2):101
Goldstein, Kenneth J. (Malcolm Pirnie Inc./USA) 5(5):87
Goltz, Mark N. (Air Force Institute of Technology/USA) 5(2):275
Gomez, Luis (Foster Wheeler Environ. Corp./USA) 5(3):131
Gonzales, James R. (U.S. Air Force/USA) 5(1):83
Gonzalez, Cristobal L. (U.S. Marine Corps/USA) 5(5):57

Gorder, Kyle (CELTIC Technologies Ltd./UK) 5(1):65
Gordon, E. Kinzie (Parsons Engineering Science, Inc./USA) 5(1):121
Gordon, Milton P. (University of Washington/USA) 5(6):133
Gosselin, Christian (Sodexen Group/ CANADA) 5(8):1
Gossett, James M. (Cornell University/ USA) 5(2):27
Gourdon, Remy P. (Institut National des Sciences Appliquees/FRANCE) 5(8):235
Govind, Rakesh (University of Cincinnati/USA) 5(4):37
Graham, David W. (University of Kansas/USA) 5(1):283
Graham, William H. (University of Kansas/USA) 5(1):283
Granade, Steve (U.S. Navy/USA) 5(2):165
Gravel, Marie-Julie (École Polytechnique de Montréal/CANADA) 5(5):135
Graves, Duane (IT Corporation/USA) 5(1):141
Gray, A.L. (Zeneca Corp./CANADA) 5(7):125
Gray, Julian (Integrated Science & Technology, Inc./USA) 5(1):97
Gray, Neil C.C. (Zeneca Corp./ CANADA) 5(7):125
Green, Roger B. (Waste Management, Inc./USA) 5(7):39, 45
Greene, Mark R. (Evirogen, Inc./USA) 5(7):7
Greenwald, Robert M. (HSI GeoTrans, Inc./USA) 5(6):169
Gregory, Betsy (The University of Tennessee/USA) 5(7):155, 161
Gregory, G.E. (URS Greiner Woodward-Clyde/USA) 5(1):35
Greis, Rolf (Institut fur Biologische Sanierung GmbH/GERMANY) 5(3):517
Griffin, Gerv C. (EA Engineering Science & Technology, Inc./USA) 5(6):251
Griffioen, Jasper (TNO Institute of Applied Geoscience/NETHERLANDS) 5(1):189; 5(3):463, 481
Griffiths, Evan (University of Idaho/ USA) 5(1):1
Grishchenkov, Vladimir G. (Russian Academy of Sciences/RUSSIA) 5(3):209
Groenwold, J. (AB-DLO/ NETHERLANDS) 5(3):529
Grøn, Christian (Risø National Laboratory/DENMARK) 5(6):15
Grossman, Matthew J. (Exxon Research & Engineering Co,/USA) 5(3):227
Grotenhuis, Tim C. (Wageningen Agricultural University/ NETHERLANDS) 5(3):195, 529; 5(8):69, 241, 265
Groudev, Stoyan N. (University of Mining & Geology/BULGARIA) 5(4):103; 5(5):7
Groudeva, Veneta I. (University of Sofia/ BULGARIA) 5(5):7
Grundl, Timothy J. (University of Wisconsin-Milwaukee/USA) 5(1):253
Guénette, Chantal C. (SINTEF Applied Chemistry/NORWAY) 5(3):227
Guérin, Valérie A.E. (Laboratoire de Géomécanique/FRANCE) 5(3):433
Guiler, H. (Zeneca Corp./CANADA) 5(7):125
Guiot, Serge R. (Biotechnology Research Institute/CANADA) 5(3):439; 5(7):15, 107
Gundrum, Jack (The Pennsylvania State University/USA) 5(2):237
Gusek, James (Knight Piesold, LLC/ USA) 5(6):217
Guy, Christophe (École Polytechnique de Montréal/CANADA) 5(5):135
Guy, John H. (Inland Pollution Services PR/USA) 5(3):319
Gwinn, Rosa E. (Dames & Moore, Inc./USA) 5(7):51

Haas, Patrick E. (U.S. Air Force/USA) 5(1):83, 313; 5(2):185
Haasnoot, C. (Logisticon Water Treatment/NETHERLANDS) 5(2):225
Haeseler, Frank (Institut Français du Pétrole/FRANCE) 5(8):117
Hagley, Michelle (Texaco Ltd./UK) 5(3):239

Haglund, Peter (Umea University/ SWEDEN) 5(8):135
Haider, Islam (King's College University/ UK) 5(3):363
Haines, John (U.S. EPA/USA) 5(3):245; 5(8):209
Haith, C E (Exxon Research & Engineering/USA) 5(3):227, 493
Hall, Jennifer M. (University of Aberdeen/UK) 5(4):25
Hall, William B. (URS Greiner Woodward-Clyde/USA) 5(1):35
Hallberg, Kevin B. (University of Wales, Bangor/UK) 5(6):225
Hannah, Marty (University of Alaska/ USA) 5(3):523
Hannula, Heli (Univ. Tempere/ITALY) 5(3):71
Hansen, Jerry E. (U.S. Air Force/USA) 5(1):121
Hansen, L. (U.S. Army Corps of Engineers/USA) 5(3):545
Happel, Harald (University of Applied Sciences/GERMANY) 5(4):97
Hard, Barbara C. (Umweltforschungszentrum/ GERMANY) 5(7):235
Hardisty, Paul E. (Komex Europe/UK) 5(6):289; 5(8):123
Hardy, Leslie (Morrow Environmental Consultants/CANADA) 5(1):19, 147
Harju, John A. (Gas Research Institute/ USA) 5(2):113
Harmsen, Joop (SC-DLO/ NETHER-LANDS) 5(6):39; 5(8):57, 87
Harold, Paul (CELTIC Technologies Ltd./UK) 5(8):63
Harre, Bryan (U.S. Navy/USA) 5(2):165
Harrington-Baker, Mary Ann (MSE Technology Applications, Inc./USA) 5(6):205
Hartley, James D. (CH2M Hill/USA) 5(3):77
Hartog, Niels (Utrecht University/ NETHERLANDS) 5(3):463
Harvey, Gregory J. (U.S. Air Force/USA) 5(6):239
Hashimoto, Akiko (Nat'l Inst for Environ. Studies & CREST/ JAPAN) 5(2):89
Hashimoto, Manabu (Towa Kagaku Co., Ltd./JAPAN) 5(2):101
Hater, Gary R. (Waste Management Inc./ USA) 5(7):39, 45
Hatzinger, Paul (Envirogen Inc./USA) 5(5):87
Hauer, Kendall L. (Dames & Moore/ USA) 5(1):271
Hawari, Jalal (National Research Council of Canada/CANADA) 5(7):15
Hayes, Thomas D. (Gas Research Institute/USA) 5(8):147
Hayes Martin, Elizabeth M. (ENSR/ USA) 5(1):247
Hazen, Terry C. (University of California/USA) 5(2):107
Heath, David (Texaco Ltd./UK) 5(3):239
Hedlund, Brian (University of Washington/USA) 5(8):191
Heerenklage, Jorn (Technische Universität Hamburg-Harburg/ GERMANY) 5(5):31
Heersche, Jan (BION Overijssel B.V/ NETHERLANDS) 5(8):87
Heimovaara, Timo J. (IWACO/ NETHERLANDS) 5(1):355
Henderson, Roger (U.S. Army Corps of Engineers/USA) 5(3):301
Henkler, Christoph (King's College University/UK) 5(3):363, 517
Henkler, Rolf D. (ICI Paints/UK) 5(3):369, 517
Henry, Bruce M. (Parsons Engineering Science, Inc/USA) 5(1):121
Henssen, Maurice J.C. (Bioclear Environmental Biotechnology/ NETHERLANDS) 5(2):225
Herbert, Susan (Environment Agency/ UK) 5(6):283
Herrmann, Ronald F. (U.S. EPA/USA) 5(8):7, 31, 209
Hernandez, Guillermina (Institut Français du Pétrole/FRANCE) 5(3):25
Hernandez-Rodríguez, Vanessa (Universidad Nacional Autonoma de Mexico/MEXICO) 5(3):409
Hers, Ian (Golder Associates Ltd. & University of British Columbia/ CANADA) 5(1):231

Herwig, Russell P. (University of Washington/USA) 5(8):315
Hess, Thomas F. (University of Idaho/ USA) 5(7):211
Hetterschijt, R. (TNO Institute of Applied Geoscience/NETHERLANDS) 5(2):141
Hickman, Randy (Key Environmental Inc./USA) 5(1):289
Hicks, James E. (Battelle/USA) 5(2):7
Hicks, Patrick H. (Regenesis Bioremediation Products/USA) 5(3):277, 289; 5(7):223
Hiebert, Randy (MSE Technology Applications/USA) 5(4):31
Hinchee, Robert E. (Parsons Engineering Science, Inc./USA) 5(1):121; 5(5):183
Hinton, Steve (Exxon Research & Engineering/USA) 5(8):209
Hintze, H. (Technical University of Hamburg-Harburg/GERMANY) 5(5):25
Hoehamer, Chris (University of Georgia/ USA) 5(6):151
Hoffmann, Robert E. (Chevron Canada Resources/CANADA) 5(5):189
Hofstra, Simon (IWACO/NETHERLANDS) 5(2):141
Holden, Robert W. (Purdue University/ USA) 5(2):47
Holder, Edith L. (University of Cincinnati/USA) 5(3):245
Holtom, Graham J. (AEA Technology/ UK) 5(3):227
Hong, Jihua (U.S. EPA/USA) 5(4):135
Hooda, I.S. (CCS Haryana Agricultural University/INDIA) 5(7):119
Horinouchi, Masae (The University of Tokyo/JAPAN) 5(8):155
Hosler, Kevin R. (Conor Pacific-WTI/ CANADA) 5(6):27
Houck, Christian E. (ThermoRetec/USA) 5(3):137
Howe, Frances (Connolly Environmental/ AUSTRALIA) 5(8):43
Huang, Jungi (Air Force Institute of Technology/USA) 5(2):275
Hubach, Cor (DHV Noord Nederland/ NETHERLANDS) 5(2):225
Hughes, Chris P. (City of Owen Sound/ CANADA) 5(6):259
Hughes, Joseph B. (Rice University/USA) 5(2):185, 287; 5(6):121
Hughes, William D. (Parsons Engineering Science, Inc./USA) 5(2):1; 5(3):283
Hullman, Aaron S. (Integrated Science & Technology/USA) 5(1):97, 219; 5(5):183
Hunter, William J. (U.S. Dept of Agriculture/USA) 5(4):47, 53
Hupe, Karsten (Consultants for Waste Management/GERMANY) 5(5):31
Hurt, Kelly L. (Dynamac Corporation/ USA) 5(1):103
Huston, Nancy (Nicor Gas/USA) 5(8):147
Hüttmann, Stephan (Groth & Co./ GERMANY) 5(3):325, 487
Hwu, Ching-Shyung (Fen-Chia University/TAIWAN - R.O.C.) 5(3):529
Hyman, Michael R. (North Carolina State University/USA) 5(3):7; 5(7):131

Iacondini, Antonella (Montecatini Environmental Research Center/ ITALY) 5(3):71
Imano, Satoshi (Asano Engineering Co, Ltd/JAPAN) 5(2):95; 5(3):233
Infante, Carmen (PDVSA Intevep/ VENEZUELA) 5(3):215
Iqbal, Muhammad Javed (King's College/UK) 5(3):535
Ivanova, Ilyana (University of Sofia/ BULGARIA) 5(5):7
Iwasaki, Kazuhiro (National Institute for Environmental Studies & CREST/ JAPAN) 5(2):89, 95, 101; 5(3):233

Jackman, Simon A. (University of Oxford/UK) 5(4):127
Jacko, Robert B. (Purdue University/ USA) 5(2):47
Jahn, Yvonne (UFZ Centre for Environmental Research/GERMANY) 5(7):143
Jankowski, Michael D. (Texas A&M University/USA) 5(7):21

Jayaraman, Vinod (North Carolina A&T State University/USA) 5(3):403
Jayaweera, Gamani R. (University of California/USA) 5(4):141
Jensen, Erik S. (Royal Agricultural University/DENMARK) 5(6):15
Jerger, Douglas E. (IT Corporation/USA) 5(7):33, 63
Jin, Bo (University of New England/USA) 5(5):203
Johns, Christopher A. (University of Alberta/CANADA) 5(1):195
Johnson, Christian D.(Battelle PNWD/USA) 5(1):29; 5(2):165
Johnson, D. Barrie (University of Wales, Bangor/UK) 5(6):225
Johnson, Michael (MSE Technology Applications, Inc./USA) 5(6):205
Johnson, Paul C. (Arizona State University/USA) 5(3):37, 123
Johnson, Richard L. (Oregon Graduate Institute/USA) 5(1):219; 5(3):123
Jones, Alexander P. (Environmental Partners, Inc./USA) 5(3):137
Jones, Antony D.G. (King's College/UK) 5(8):123
Jones, D. Martin (University of Newcastle-upon-Tyne/UK) 5(3):239, 251
Jones, David (Shell Research and Technology Centre Thornton/UK) 5(3):239
Jones, Kelly A. (AGI Technologies/USA) 5(1):331
Jones, Kevin C. (Lancaster University/UK) 5(8):25, 253
Jones, Sonya A. (U.S. Geological Survey/USA) 5(6):101
Jones, W. Jack (U.S. EPA/USA) 5(7):95
Jonioh, Viola (National University of Singapore/SINGAPORE) 5(4):7
Jowett, E. Craig (Waterloo Biofilter Systems Inc./CANADA) 5(6):259
Juntongjin, Kanchana (Chulalongkorn University) 5(8):155
Just, Eric M. (Oak Ridge National Laboratory/USA) 5(7):185
Jütterschenke, Peter (ARBES-Umwelt GmbH/GERMANY) 5(3):487
Kallur, Shiva (Environmental Strategies & Applications/USA) 5(2):181
Kamel, Ramey (University of Southern California/USA) 5(6):175
Kampf, Ruud (Uiterwaterende Sluizen/NETHERLANDS) 5(6):39
Kamphausen, Stefan (Düsseldorfer Consult GMBH/GERMANY) 5(8):141
Kastner, James R. (University of Georgia/USA) 5(2):191
Kearney, Theresa (Environment Agency/UK) 5(6):283
Keim, Lois G. (Exxon Research & Engineering/USA) 5(3):227
Kelley, Sara L. (University of Iowa/USA) 5(6):139
Kelly, David S. (IT Corporation/USA) 5(3):301
Kemper, Mark (U.S. EPA/USA) 5(2):263
Keuning, Sytze (Bioclear Environmental Biotechnology B V/NETHERLANDS) 5(2):225
Kiilerich, Ole (HOH Water Technology/DENMARK) 5(2):81
Killham, Ken (University of Aberdeen/UK) 5(3):363, 369; 5(4):25; 5(7):247
Kilmer, Kenneth W. (EA Engineering Science & Technology/USA) 5(6):251
Kim, Han S. (The University of Michigan/USA) 5(6):75
Kim, Kijung (The Pennsylvania State University/USA) 5(4):147
Kim, Soochul (Korea Institute of Science & Technology/REP OF KOREA) 5(6):245
Kim, Y.U. (The Pennsylvania State University/USA) 5(2):237
Kinal, Rebecca (General Physics Corporation/USA) 5(3):115
King, Mark (Groundwater Insight/CANADA) 5(1):289
Kinney, Kerry A. (The University of Texas at Austin/USA) 5(5):117, 123
Kinsall, Barry L. (Oak Ridge National Laboratory/USA) 5(1):47
Kivanç, Sedat (Middle East Technical University/TURKEY) 5(5):81

Klasson, K. Thomas (Oak Ridge National Laboratory/USA) 5(7):185
Klecka, Gary M. (The Dow Chemical Company/USA) 5(1):29
Klens, Julia L. (IT Corporation/USA) 5(1):141
Klint, Steen (Simcoe Co./CANADA) 5(6):259
Knightes, Christopher D. (Princeton University/USA) 5(8):173
Knowles, Christopher J. (University of Oxford/UK) 5(4):127
Kodikanti, Madhumohand (Texas A&M University/USA) 5(7):21
Koenen, Brent A. (USA-CRREL, AK/USA) 5(6):33
Koenigsberg, Stephen S. (Regenesis Bioremediation Products/USA) 5(2):61, 67, 121, 181, 287; 5(3):13; 5(7):223
Kokwaro, J.O. (University of Nairobi/KENYA) 5(7):69
Kolkman, Iris (Delft Geotechnics/NETHERLANDS) 5(3):109
Kolmert, Asa K. (University of Wales/UK) 5(6):225
Komnitsas, Kostas (University of Athens/GREECE) 5(4):103
Koning, Michael (Technische University of Hamburg-Harburg/GERMANY) 5(5):25, 43
Koons, Brad W. (University of Minnesota/USA) 5(2):41
Korbini, Ron (Investment Recovery LLC./USA) 5(1):289
Kosar, K.M (EBA Engineering Consultants Ltd./CANADA) 5(2):249
Kosson, David S. (Rutgers University/USA) 5(8):271
Kotterman, Michiel J.J. (Wageningen Agricultural University/NETHERLANDS) 5(8):69
Koutsky, Bouhmil (Institute of Chemical Technology/CZECH REPUBLIC) 5(5):161
Krieger-Brockett, Barbara (University of Washington/USA) 5(8):301
Krishnayya, Addanki V. (O'Connor Associates Environmental Inc./CANADA) 5(5):111
Kronenburg, Rene (Gemeentelijk Havenbedrijf Amsterdam/NETHERLANDS) 5(3):395
Krueger, Jennifer J. (Thomas Consultants, Inc./USA) 5(1):271; 5(6):21
Kryczkowski, Rosann (ITT Industries Night Vision/USA) 5(2):255
Kuan, H.Ling (University of Aberdeen/UK) 5(3):369
Kubal, Jerry (Kubal-Furr & Associates/USA) 5(7):223
Kubatova, Alena (Academy of Sciences of the Czech Republic/CZECH REPUBLIC) 5(6):69
Kubota, Katsuyuki (Towa Kagaku Co., Ltd./JAPAN) 5(2):101
Kuhn, Mark (Hydro Geo Chem, Inc./USA) 5(3):59
Kumagai, M. (Lake Biwa Research Institute/JAPAN) 5(8):309
Kuniansky, Eve L. (U.S. Geological Survey/USA) 5(6):101
Kurrus, Joseph A. (AVISTA Corporation/USA) 5(3):189
Kusaka, Jun (Gunma University/JAPAN) 5(4):73

Lahr, Mark (Key Environmental Inc./USA) 5(1):289
Lakshmiprasad, Thimappa (Utah State University/USA) 5(3):183
Lam, Sylvia (Simon Fraser University/CANADA) 5(8):75
Lamar, Richard T. (EarthFax Development Corp/USA) 5(8):81
Lamb, Steven R. (GZA GeoEnvironmental, Inc./USA) 5(2):171
Lambert, Patrick (Environment Canada/CANADA) 5(3):227
Langenhoff, Alette (TNO Institute of Environmental Science/NETHERLANDS) 5(3):481; 5(6):81

Langwaldt, Jörg H. (Tampere University of Technology/FINLAND) 5(7):113
Laturnus, Frank (Risø National Laboratory/DENMARK) 5(6):15
Lau, Sok-Kiang (DSO National Laboratories/SINGAPORE) 5(3):257
Launen, Loren A. (Simon Fraser University/CANADA) 5(1):147; 5(8):75
Lawson, A. (Environment Canada/ CANADA) 5(8):309
Layton, Alice (The University of Tennessee/USA) 5(7):155
Le Roux, Françoise (Institut Français du Pétrole/FRANCE) 5(3):25
Leal, P. (Campinas State University-UNICAMP/BRAZIL) 5(3):457
Lebeault, Jean-Michel (Université de Technologie de Compiègne/FRANCE) 5(1):177
Lee, Annette (XDD, LLC/USA) 5(6):1
Lee, Chi Mei (National Chung Hsing University/TAIWAN - R.O.C.) 5(3):469
Lee, Kenneth (Fisheries & Oceans Canada/CANADA) 5(3):227, 239
Lee, Michael D. (Terra Systems, Inc./ USA) 5(2):55
Lee, Philip H. (Lancaster University/UK) 5(8):259
Lee, Roger W. (U.S. Geological Survey/ USA) 5(6):101
Leeson, Andrea (Battelle/USA) 5(3):89
Legrand, Robert (Radian International LLC/USA) 5(2):113
Leigh, Daniel P. (IT Group, Inc./USA) 5(2):165
Lemons, Ken (Black & Veatch/USA) 5(3):289
Lemos, Barbara B. (Earth Tech/USA) 5(2):255
Lerner, David N. (University of Sheffield/UK) 5(1):225, 277
Lesage, Suzanne (National Water Research Institute/CANADA) 5(8):161
Lessard, Drew F. (U.S. Air Force/USA) 5(3):163
Lethbridge, Gordon (Shell Research and Technology Centre Thornton/UK) 5(3):239
Lettinga, Gatze (Agricultural University Wageningen/NETHERLANDS) 5(7):253
Levesque, Marie-Josee (Biotechnology Research Institute/CANADA) 5(7):107
Li, Loretta (University of British Columbia/CANADA) 5(1):231
Li, Wan-Chi (National Water Research Institute/CANADA) 5(8):161
Linhares, Monica M. (Petrobras/ CENPES/BRAZIL) 5(3):421
Linich, Michael A. (University of Newcastle/AUSTRALIA) 5(6):277
Liou, Raycharn (Howard University/ USA) 5(7):179
Litchfield, Carol D. (George Mason University/USA) 5(5):143
Litherland, Susan Tighe (Roy F. Weston, Inc./USA) 5(2):157
Liu, Chung-Chiun (Case Western Reserve University/USA) 5(1):331
Liu, Dickson (National Water Research Institute/CANADA) 5(8):161
Liu, Shung-Hung (National Chung Hsing University/TAIWAN – R.O.C.) 5(3):469
Lizotte, Craig (Envirogen Inc./USA) 5(6):1
Loeffler, Frank E. (Michigan State University/USA) 5(2):21
Logan, Bruce E. (The Pennsylvania State University/USA) 5(2):231; 5(4):147
Logan, Carla M. (Baltimore Gas & Electric Co./USA) 5(1):295
Lorenzen, L. (University of Stellenbosch/REP. OF SOUTH AFRICA) 5(5):51
Losi, Mark E. (Foster Wheeler Environ. Corp./USA) 5(3):383
Love, Nancy G. (Virginia Tech./USA) 5(8):19
Lovley, Derek R. (University of Massachusetts/USA) 5(1):331
Lowder, Anna (U.S. Army/USA) 5(7):1

Lu, Chih-Jen (National Chung Hsing University/TAIWAN – R.O.C.) 5(2):205; 5(3):469
Lu, Yongming (Biospherics Incorporated/USA) 5(4):65
Ludvigsen, Liselotte (HOH Water Technology/DENMARK) 5(2):81
Ludwig, Ralph (U.S. EPA/USA) 5(4):19
Lundanes, Elsa (University of Oslo/NORWAY) 5(6):187
Lundstedt, Staffan (University of Umea/SWEDEN) 5(8):135
Luth, Joachim C. (Technical University of Hamburg-Harburg/GERMANY) 5(5):25, 43
Luthy, Richard G. (Carnegie Mellon University/USA) 5(7):241; 5(8):289
Lutze, Werner (University of New Mexico/USA) 5(4):59, 65, 115
Lysebo, Carina (University of Oslo/NORWAY) 5(6):187

Ma, Wim (IBN-DLO/NETHERLANDS) 5(6):39
MacFarlane, Ian D. (EA Engineering Science & Technology, Inc./USA) 5(1):295; 5(6):251
Mack, E. Erin (National Research Council/USA) 5(7):95
MacLeod, Christopher J.A. (Lancaster University/UK) 5(6):109; 5(8):217, 259
Macnaughton, Sarah J. (University of Tennessee/USA) 5(3):345; 5(4):1
Magaña, Miguel (Universidad Autonoma Metropolitana-Iztapalapa/MEXICO) 5(3):31
Mage, Roland (Battelle Europe/SWITZERLAND) 5(1):159
Mahaffey, William R. (Pelorus Environmental Biotech Corp./USA) 5(1):71; 5(2):293; 5(3):13; 5(5):19
Mahjoub, Borhane (Institut National des Sciences Appliquees/FRANCE) 5(8):235
Maini, Giacomo (IBS Viridian Ltd./UK) 5(4):127
Malaspina, F. (Montecatini Environmental Research Center/ITALY) 5(3):71
Malina, Grzegorz (Technical University of Czestochowa/POLAND) 5(3):195
Man, Malcolm K. (Morrow Environmental Consultants Inc./CANADA) 5(1):147
Manchester, Kenneth (MSE Technology Applications, Inc./USA) 5(6):205
Mandava, Shanthi (Earth Tech/USA) 5(1):241; 5(3):289
Mandel, Robert (U.S. EPA/USA) 5(6):89
Maner, Paul (Equilon Enterprises LLC [Shell/Texaco]/USA) 5(3):37
Marchal, Rémy (Institut Français du Pétrole/FRANCE) 5(1):177
Marcon, Michael F. (InControl Technologies/USA) 5(1):7
Marley, Michael C. (XDD, LLC/USA) 5(6):1
Marnette, Emile C.L.(Tauw Milieu bv/NETHERLANDS) 5(3):83, 109
Márquez-Rocha, Facundo J. (Centro Invest Cient y Educ Super (CICESE)/MEXICO) 5(3):409
Martin, Geoffrey R. (University of Southern California/USA) 5(6):175
Martin, Ian (Environment Agency/UK) 5(6):283
Martineau, Geneviève (École Polytechnique de Montréal/CANADA) 5(3):451
Martínez-Cruz, J. (CINVESTAV-IPN/MEXICO) 5(7):149
Mason, Anna (Beatty Franz & Associates/CANADA) 5(2):281
Mason, Jeremy R. (King's College/UK) 5(3):363, 535; 5(8):111, 123
Matsubara, Takashi (Obayashi Corporation/JAPAN) 5(5):13
Mazur, Chris (U.S. EPA/USA) 5(6):151
Mazur, Margaret (Connolly Environmental/AUSTRALIA) 5(8):43
Mazzanti, Mariangela (Centro Ricerche Ambientali Montecatini/ITALY) 5(3):511
McBean, Edward A. (Conestoga-Rovers & Associates/CANADA) 5(1):77, 325
McCall, Sarah (Battelle/USA) 5(2):7

McCall, Wesley (Geoprobe Systems/ USA) 5(1):337
McCauley, Paul T. (U.S. EPA/USA) 5(8):7
McClellan, Kim (U.S. EPA/USA) 5(8):209
McCutcheon, Steven (U.S. EPA/USA) 5(6):239
McDermott, Raymond (U.S. Army/ USA) 5(3):115
McDermott, Tanya (Mongomery Watson/USA) 5(1):41
McDonald, Thomas J. (Texas A&M University/USA) 5(3):209
McGrath, Steve (Rothamsted Experimental Station/UK) 5(4):25
McGregor, Rick (Conor Pacific-WTI/ CANADA) 5(4):19
McIntosh, N.D. Philip (The University of Texas/USA) 5(5):123
McInturff, Frank (EnSafe, Inc./USA) 5(8):37
McKay, Daniel J. (U.S. Army/USA) 5(3):169
McLean, Jeff S. (University of Guelph/ CANADA) 5(4):121
McLinn, Eugene L. (RMT, Inc./USA) 5(1):183
McMonagle, Thomas L. (EarthTech/USA) 5(3):143
McNamara, Sean (Carnegie Mellon University/USA) 5(7):241
Meharg, A.A. (Institute of Terrestrial Ecology/UK) 5(7):247
Mehran, Mohsen (Geosystem Consultants, Inc./USA) 5(1):13
Meijerink, E. (Province of Drenthe/ NETHERLANDS) 5(2):225
Melzer, P. (Mannheimer Versorgungs und Verkehrsgesellschaft mbH (MVV)/GERMANY) 5(3):337
Menaia, Jose (Instituto Nacional de Engenharia e Tecnologia Industrial/ PORTUGAL) 5(5):197
Menn, Fu-Min (The University of Tennessee/USA) 5(7):155, 161
Merckx, Roel (Catholic University Leuven/BELGIUM) 5(8):179
Merli, Carlo (Universita di Roma La Sapienza/ITALY) 5(6):157
Mesarch, Matthew (Purdue University/ USA) 5(3):357
Metris, Aline V.(University of Teesside/ UK) 5(5):161
Michelberger, H.S. (TAUW Milieu bv./NETHERLANDS) 5(3):445
Mickel, Charles E. (SECOR International Inc./USA) 5(1):71
Miksch, Korneliusz (Silesian Technical University/POLAND) 5(8):203
Millar, Kelly (National Water Research Institute/CANADA) 5(8):161
Miller, Aaron (Doe Run Co/USA) 5(6):217
Miller, David (Sandia National Laboratories/USA) 5(1):89
Miller, Joel (The Pennsylvania State University/USA) 5(4):147
Miller, Karen M. (University of Cincinnati/USA) 5(3):245
Miller, Michael E. (Camp Dresser & McKee, Inc./USA) 5(1):259
Miller, T. Ferrell (Roy F. Weston, Inc./REAC/USA) 5(6):89
Millette, Denis (Biotechnology Research Institute, NRC/CANADA) 5(3):177
Mills, DeEtta K. (George Mason University/USA) 5(5):143
Milne, Roger (Environment Agency/UK) 5(3):239
Milner, Larry (Black & Veatch Waste Science, Inc./USA) 5(8):147
Minsker, Barbara S. (University of Illinois/USA) 5(1):301, 349
Mitchell, David J. (AEA Technology plc/UK) 5(3):239
Mobarry, Bruce K. (University of Idaho/ USA) 5(1):1
Moeri, Ernesto N. (CSD-Geoklock/ BRAZIL) 5(1):19
Molina, Marirosa (U.S. EPA/USA) 5(2):191
Molitoris, Hans Peter (University of Regensburg/GERMANY) 5(6):69
Momeni, Dawood (University of Southern California/USA) 5(6):175

Montalvo, Claudia (CINVESTAV-IPN/MEXICO) 5(8):93

Montgomery, Michael T. (U.S. Navy/USA) 5(8):223, 295

Moore, Margo M. (Simon Fraser University/CANADA) 5(1):147; 5(8):75

Morecraft, Andrew J. (Radian International LLC/USA) 5(2):113

Morgan, Philip (ICI Technology/UK) 5(2):281

Morin, Thomas (Environmental Partners, Inc./USA) 5(3):137

Morley, Matthew C. (The University of Texas at Austin/USA) 5(7):217

Morley, Theresa (U.S. Navy/USA) 5(3):131, 383

Mortensen, Gerda K. (Risø National Laboratory/DENMARK) 5(6):15

Moser, G.P. (Zeneca Corp./CANADA) 5(7):125

Moser, Lori E. (Zeneca Corp./CANADA) 5(2):281; 5(7):125

Mourik, J. (Logisticon Water Treatment/NETHERLANDS) 5(2):225

Mulder, G.J. (Delft Geotechnics/NETHERLANDS) 5(3):103

Müller, Bernd G. (National Research Center for Biotechnology GBF/GERMANY) 5(5):155

Müller, Roland A. (UFZ Centre for Environmental Research/GERMANY) 5(7):143

Müller, Roland H. (UFZ Centre for Environmental Research/GERMANY) 5(7):143

Mullick, Anjum (University of Alberta/CANADA) 5(1):195

Mulvaney, Peter (The Pennsylvania State University/USA) 5(4):147

Munakata-Marr, Junko (Colorado School of Mines/USA) 5(2):293

Munster, Clyde (Texas A&M University/USA) 5(6):9

Murali, Dev M. (General Physics Corporation/USA) 5(3):115

Murdoch, Lawrence C. (Clemson University/USA) 5(2):263

Murphy, J. Richard (Conestoga-Rovers & Associates/CANADA) 5(1):77, 325

Murphy, Tom (Environment Canada/CANADA) 5(8):309

Murray, Willard (Harding Lawson Associates/USA) 5(2):121

Myers, Michael J. (The Ohio State University/USA) 5(6):181

Mysona, Eric S. (Parsons Engineering Science, Inc./USA) 5(3):283

Mutafov, Sava (Institute of Microbiology/BULGARIA) 5(5):197

Nadeau, Royal (U.S. EPA/USA) 5(8):51

Nakasugi, Naou (National Institute for Environmental Studies & CREST/JAPAN) 5(2):89

Nakatsu, Cindy (Purdue University/USA) 5(3):357

Neaville, Chris C. (Equilon Enterprises LLC [Shell/Texaco]/USA) 5(3):37

Neff, Jerry (Battelle/USA) 5(5):69

Nehra, Ajit S. (CCS Haryana Agricultural University/INDIA) 5(7):119

Nelson, Christopher H. (IT Corp/USA) 5(2):199

Nelson, Mark S. (Williams Gas Pipeline-Transco/USA) 5(2):113

Nelson, Sheldon (Chevron Research & Technology Co/USA) 5(4):91

Newell, Charles J. (Groundwater Services, Inc./USA) 5(1):83; 5(2):185

Newman, Lee A. (University of Washington/USA) 5(6):133

Ng, Lee-Ching (DSO National Laboratories/SINGAPORE) 5(3):257

Ng, Wei-Fang (DSO National Laboratories/SINGAPORE) 5(3):257

Nichols, H.L. (University of Arkansas/USA) 5(6):33

Niederreither, Michael S. (EarthTech/USA) 5(3):143

Nielsen, Soren Toft (Scanrail Consult/DENMARK) 5(3):541

Nieman, J. Karl C. (Utah State University/USA) 5(5):97

Nies, Loring (Purdue University/USA) 5(2):47; 5(3):357

Nipshagen, A.A.M. (IWACO bv/ NETHERLANDS) 5(3):481
Nishino, Shirley (U.S. Air Force/USA) 5(7):7
Nivens, David E. (Microbial Insights/ USA) 5(8):277
Nocentini, Massimo (Universita di Bologna/ITALY) 5(3):313
Noftsker, Christina (Exponent Environmental Group/USA) 5(7):83
Nogalski, Nick A. (Envirogen, Inc./USA) 5(7):7
Nojiri, Hideaki (The University of Tokyo/JAPAN) 5(8):155
Nordwick, Suzzann (MSE Technology Applications/USA) 5(4):31
Norris, Robert D. (Eckenfelder/Brown and Caldwell/USA) 5(1):53, 59
Novak, John T. (Virginia Polytechnic Institute & State Univ/USA) 5(6):45; 5(7):191; 5(8):19
Novak, Paige J. (University of Minnesota/USA) 5(2):35, 41
Novotny, Cenek (Czech Academy of Sciences/CZECH REPUBLIC) 5(6):69
Nuttall, H. Eric (University of New Mexico/USA) 5(4):59, 65, 115
Nzengung, Valentine A. (University of Georgia/USA) 5(6):145, 239

Obbard, Jeffrey P. (National University of Singapore/SINGAPORE) 5(4):7
Öberg, Lars G. (University of Umea/ SWEDEN) 5(8):135
Obermann, Peter (University of Bochum/GERMANY) 5(3):475
Ochanda, John O. (University of Nairobi/KENYA) 5(7):69
Ochs, L. Donald (Regenesis Bioremediation Products/USA) 5(3):271; 5(6):169
Odom, J. Martin (DuPont Company/ USA) 5(1):35
Ohashi, Takashi (Asano Engineering Co, Ltd/JAPAN) 5(2):95; 5(3):233
O'Melia, Barry (URS Greiner Woodward Clyde/USA) 5(8):167
Omori, Toshio (The University of Tokyo/JAPAN) 5(8):155
O'Neill, Kathleen (O'Neill Associates/ USA) 5(8):209
Opsteeg, Joris L. (Wageningen Agricultural University/ NETHERLANDS) 5(7):253
Orchard, Brady J. (Utah State University/USA) 5(6):127
O'Reilly, Kirk T. (Chevron Research & Technology Co/USA) 5(3):7; 5(4):91
Orolin, John (EverClear Environmental Technologies/USA) 5(3):375
Osano, Anne Adhiambo (University of Nairobi/KENYA) 5(7):69
O'Shaughnessy, James C. (Worcester Polytechnic Institute/USA) 5(3):265
Oudijk, Gil (Triassic Technologies, Inc/USA) 5(3):271, 319
Owens, Edward H. (Ploaris Applied Sciences/USA) 5(3):227
Owens, Victor (Foster Wheeler Environmental/USA) 5(3):131

Paca, Jan (Institute of Chemical Technology/CZECH REPUBLIC) 5(5):161, 197
Packwood, Calvin R.X. (University of Sheffield/UK) 5(1):225
Pak, Daewon (Korea Institute of Science & Technology/REP OF KOREA) 5(6):245
Palagyi, Tony (Equiva Services/USA) 5(3):137
Palumbo, Anthony V. (Oak Ridge National Laboratory/USA) 5(1):47; 5(4):1
Pannatier, Elisabeth Graf (Netherlands Institute of Applied Geosci TNO (NITG-TNO)/NETHERLANDS) 5(1):189
Paquet, Loiuse (National Research Council of Canada/CANADA) 5(7):15
Patel, D. (The National Grid Co./UK) 5(8):259
Patel, Madhu (EnSafe/USA) 5(8):37
Paton, Graeme I. (University of Aberdeen/UK) 5(3):351; 5(4):25; 5(8):217
Pavlostathis, Spyros G. (Georgia Institute of Technology/USA) 5(7):259

Peacock, Aaron (University of Tennessee/USA) 5(3):345
Peargin, Tom (Chevron Research & Technology Co/USA) 5(3):19
Pelton, Deborah K. (U.S. Army/USA) 5(6):33
Pendrell, Doug (U.S. Army Corps of Engineers/USA) 5(3):289
Percival, Paul (Simon Fraser University/ CANADA) 5(8):75
Pérez, Sixto (Instituto Mexicano de Tecnologia del Aqua/MEXICO) 5(8):197
Perille, Mary Lynne (University of Wisconsin-Milwaukee/USA) 5(1):253
Perkins, Ed (AscI/USA) 5(3):545
Perrier, Michel (École Polytechnique de Montréal/CANADA) 5(3):177
Perry, L.B. (U.S. Army/USA) 5(6):33
Petch, Stuart (University of Newcastle-upon-Tyne/UK) 5(3):239
Peters, Catherine A. (Princeton University/USA) 5(8):173
Peters, Hauke (Groth & Co./ GERMANY) 5(3):487
Peterson, Lance N. (Lockheed Martin Idaho Technologies Co./USA) 5(2):147
Petrucci, Elisabetta (Universita degli Studi di Roma La Sapienza/ITALY) 5(6):157
Pfiffner, Susan M. (University of Tennessee/USA) 5(1):47; 5(3):345; 5(4):1
Phelps, Tommy J. (Oak Ridge National Laboratory/USA) 5(1):47
Phillips, Peter C. (Clark Atlanta University/USA) 5(4):109
Phillips, Ryan (C-CORE/CANADA) 5(2):249
Phipps, David (Total Forest Ind. Ltd./ CANADA) 5(4):121
Pickard, Michael A. (University of Alberta/CANADA) 5(5):211
Pidgeon, Chad S. (U.S. Army/USA) 5(6):33
Pijls, C.G.J.M. (TAUW Milieu/ NETHERLANDS) 5(3):83, 445; 5(6):81
Pinelli, Davide (University of Bologna/ ITALY) 5(3):313
Pinto, Linda (Simon Fraser University/ CANADA) 5(8):75
Pinza, M.R. (Battelle Marine Sciences Laboratory/USA) 5(6):63
Piroth, Klaus (Arcadis Trischler und Partner/GERMANY) 5(3):337
Plata, Nadia (Battelle Geneva/ SWITZERLAND) 5(6):51
Poggi-Varaldo, Hector M. (CINVESTAV del IPN/MEXICO) 5(6):265; 5(7):149
Pollack, Albert J. (Battelle/USA) 5(3):201; 5(5):57
Pomeroy, Martin (Environment Canada/ CANADA) 5(4):19
Porta, Augusto (Battelle Geneva/ SWITZERLAND) 5(1):159, 359; 5(5):69; 5(6):51
Potter, Carl L. (U.S. EPA/USA) 5(8):31, 209
Potter, Dennis (U.S. Army Corps of Engineers/USA) 5(7):57
Powell, Rhea (North Carolina A&T State University/USA) 5(3):403
Prentice, George L. (City of Owen Sound/CANADA) 5(6):259
Presecan, Mary (Parsons Engineering Science/USA) 5(8):13
Preston, Sara (University of Aberdeen/ UK) 5(4):25
Prince, Roger C. (Exxon Research & Engineering/USA) 5(3):227, 493
Pringle, Eric (Conor Pacific-WTI/ CANADA) 5(4):19
Prytula, Mark T. (Georgia Institute of Technology/USA) 5(7):259
Puhakka, Jaakko A. (Tampere University of Technology/FINLAND) 5(7):113
Puls, Robert W. (U.S. EPA/USA) 5(6):163

Quigley, Sean P. (University of Sheffield/UK) 5(1):307
Quinn, Kenneth J. (Montgomery Watson/USA) 5(2):61

Race, Susan L (BE&K/Terranext/USA) 5(3):295
Raetz, Richard M. (Global Remediation Technologies, Inc./USA) 5(2):217
Ramírez-Oliva, A. (Ecatepec Institute of Tech.-TESE/MEXICO) 5(6):265
Ransbottom, Daniel T. (ENTRIX, Inc./USA) 5(3):307
Rattray, E.A.S. (University of Aberdeen/ UK) 5(4):25
Rawal, Bhavin (University of Regensburg/GERMANY) 5(6):69
Redman, J. David (University of Waterloo/CANADA) 5(6):259
Reed, Patrick (University of Illinois/ USA) 5(1):301
Rees, Dafydd (CELTIC Technologies Ltd./UK) 5(1):65
Rees, John F. (CELTIC Technologies, Ltd./UK) 5(1):65; 5(8):63
Reid, Barry (Integrated Science & Technology/USA) 5(1):97
Reid, Brian J. (Lancaster University/ UK) 5(6):109; 5(8):25, 217, 247, 253, 259
Reid, James J. (ARCADIS Geraghty & Miller, Inc./USA) 5(2):135
Reifenstuhl, R. (Technical University of Hamburg-Harburg/GERMANY) 5(5):25
Reiners, Michael (Universität Hoxter/ GERMANY) 5(8):141
Reisinger, H. James (Integrated Science & Technology Inc/USA) 5(1):97, 219; 5(3):131; 5(5):183
Renn, Tim (University of Idaho/USA) 5(7):211
Revah, Sergio (Universidad Autonoma Metropolitana-Iztapalapa/MEXICO) 5(3):31
Reynolds, Charles M. (U.S. Army/USA) 5(6):33
Rhykerd, Robert L. (Texas A&M University/USA) 5(6):9
Rich, Peter (HSI GeoTrans, Inc./USA) 5(6):169
Riedstra, Durk (BION Overijssel B.V./ NETHERLANDS) 5(8):87
Riefler, R. Guy (University of Connecticut/USA) 5(7):27
Rijnaarts, Huub H.M. (TNO Institute of Environmental Science/ NETHERLANDS) 5(2):73; 5(3):481; 5(6):81
Rike, Anne Gunn (Norwegian Geotechnical Institute/NORWAY) 5(6):187
Ringelberg, David (Dyntel Corp./USA) 5(3):545
Ripp, Steven (University of Tennessee/ USA) 5(8):277
Riznychok, Warren M. (Malcolm Pirnie, Inc./USA) 5(7):7
Rizzuto, David L. (Indian Environmental/USA) 5(3):163
Robb, Joseph (ENSR/USA) 5(1):247
Roberge, François (École Polytechnique de Montréal/CANADA) 5(5):135
Robertiello, A (EniTecnologie/ITALY) 5(8):129
Roberts, Greg (IT Corporation/USA) 5(1):141
Robold, Eberhard (Arcadis Trischler und Partner/GERMANY) 5(3):337
Rockne, Karl J. (Rutgers University/ USA) 5(8):191, 271
Rodríguez, Elisabeth (Universidad Nacional Autonoma de Mexico/ MEXICO) 5(5):211
Rodríguez-Vázquez, Refugio (CINVESTAV-IPN/MEXICO) 5(7):149, 173; 5(8):93
Rogers, Brock (URS Greiner Woodward Clyde/USA) 5(2):231
Rohns, Hans-Peter (Stadtwerke Düsseldorf AG/GERMANY) 5(3):475; 5(8):141
Rohricht, Markus (University of Applied Sciences/GERMANY) 5(4):97
Rojas-Avelizapa, Norma G. (CINVESTAV-IPN/MEXICO) 5(7):149
Romero, Miguel (PDVSA Interep/ VENEZUELA) 5(3):215
Ross, Nathalie (École Polytechnique de Montréal/CANADA) 5(2):243

Roulier, Michael H.(U.S. EPA/USA) 5(2):263
Rovers, Frank A. (Conestoga-Rovers & Associates/CANADA) 5(1):77, 325
Ruiz-Aguilar, Graciela (CINVESTAV-IPN/MEXICO) 5(7):173
Ruiz-Saucedo, U. (Technical University of Hamburg-Harburg/GERMANY) 5(5):43
Rulkens, Wim H. (Wageningen Agricultural University/ NETHERLANDS) 5(3):195; 5(8):241, 265
Rybka, Andrzej (EXBUD-Hydrogeotechnics/POLAND) 5(5):1

Saber, Diane L. (Institute of Gas Technology/USA) 5(8):147
Sada, Ebenezar (Howard University/ USA) 5(7):179
Sadeghipour, Jamshid (Foster Wheeler/ USA) 5(3):131, 383
Sahagun, Tracy (U.S. Marine Corps/ USA) 5(3):89
Sakakibara, Yutaka (Gunma University/ JAPAN) 5(4):73
Salanitro, Joseph P. (Equilon Enterprises LLC [Shell/Texaco]/USA) 5(3):37
Salpas, Peter (SAIC/USA) 5(1):47
Salvador, Maria Cristina (CSD-Geoklock/BRAZIL) 5(1):19
Sampsel, Eric R. (University of Tennessee/USA) 5(7):155, 161
Samson, Réjean (École Polytechnique de Montréal/CANADA) 5(2):243; 5(3):177, 451; 5(5):63, 135; 5(6):211; 5(7):89, 229
Sandefur, Craig A. (Regenesis Bioremediation Products/USA) 5(2):61; 5(3):13, 277
Sanford, Robert A. (University of Illinois/USA) 5(2):21; 5(8):191
Sanseverino, John (University of Tennessee/USA) 5(7):155, 161
Santa Anna, Lidia M. (Petrobras/ CENPES/BRAZIL) 5(3):421
Santacruz, Germán (Instituto Mexicano de Tecnologia del Agua/MEXICO) 5(7):137
Santangelo-Dreiling, Theresa (Colorado Dept. of Transportation/USA) 5(1):71
Santini, Katia (École Polytechnique de Montréal/CANADA) 5(8):185
Santo Domingo, Jorge W. (U.S. EPA/ USA) 5(2):191
Sargent, Stacy (EarthTech/USA) 5(1):241
Šašek, Václav (Czechoslovak Academy of Sciences/CZECH REPUBLIC) 5(6):69
Sauerland, Uwe (University of Applied Sciences/GERMANY) 5(4):97
Sayler, Gary S. (The University of Tennessee/USA) 5(7):155, 161; 5(8):277
Sayles, Gregory D. (U.S. EPA/USA) 5(2):281
Schaar, Ralph (Malcolm Pirnie Inc./ USA) 5(5):87
Schaffner, I. Richard (GZA GeoEnvironmental, Inc./USA) 5(2):171
Schettino-Bermúdez, B. (Ecatepec Institute of Tech.-TESE/MEXICO) 5(6):265
Schimmel, Keith (North Carolina A&T State University/USA) 5(3):403
Schmauder, Hans-Peter (Forschungs-zentr fur Medizintech & Biotech e.V./GERMANY) 5(5):161, 197
Schnoor, Jerald L. (University of Iowa/ USA) 5(6):139
Schoefs, Olivier (École Polytechnique de Montréal/CANADA) 5(5):63
Scholz, Jürgen (Technical University of Braunschweig/GERMANY) 5(5):155
Schraa, Gosse (Wageningen Agricultural University/NETHERLANDS) 5(6):81
Schrader, Paul (University of Idaho/ USA) 5(7):211
Schramedei, Ralf (Düsseldorfer Consult GMBH/GERMANY) 5(8):141
Schryver, Jack C. (Oak Ridge National Laboratory/USA) 5(4):1

Schubert, Jürgen (Stadtwerke Düsseldorf AG/GERMANY) 5(3):475; 5(8):141
Schultz, T. Wayne (The University of Tennessee/USA) 5(7):155
Schulz-Berendt, Volker (Umweltschutz Nord GmbH & Co/GERMANY) 5(5):43
Schwedes, Jörg (Technical University of Braunschweig/GERMANY) 5(5):155
Scrudato, Ronald J. (State University of New York at Oswego/USA) 5(7):167
Seabra, Paulo Negrais (Petrobras/CENPES/BRAZIL) 5(3):421
Seagren, Eric A. (University of Maryland/USA) 5(1):343
Seki, Naoki (Ebara Corporation/JAPAN) 5(2):129
Semple, Kirk T. (Lancaster University/UK) 5(6):109; 5(8):25, 217, 247, 253, 259
Senrick, Maira (Watervliet Arsenal/USA) 5(5):87
Sergy, G.A. (Environment Canada/CANADA) 5(3):227
Serra, Roberto (Centro Ricerche Ambientali Montecatini/ITALY) 5(3):71, 511
Severn, Shawn R.T. (Exponent Environmental Group/USA) 5(7):83
Sewell, Guy W. (U.S. EPA/USA) 5(4):135
Sextro, Robert K. (Jacobs Engineering/USA) 5(1):153
Shammas, Samer (The University of Texas at Austin/USA) 5(7):217
Shan, Jun (University of Nebraska-Lincoln/USA) 5(4):85
Shanks, Jacqueline Vanni (Rice University/USA) 5(6):121
Sharman, Ajay (IBS Viridian Ltd./UK) 5(4):127
Sharp, Robert R. (Manhattan College/USA) 5(2):211
Shaw, Elizabeth J. (Institute of Terrestrial Ecology/UK) 5(7):247
Shaw, Travis C. (U.S. Army Corps of Engineers/USA) 5(8):315
Sheldon, John K. (Montgomery Watson/USA) 5(2):61
Shen, Chun-Fang (National Research Council of Canada/CANADA) 5(7):15
Shen, Hai (Dynamac Corporation/USA) 5(1):115
Shi, Zhou (University of Tennessee/USA) 5(7):155, 161
Shields, Adrian (Komex Clark Bond/UK) 5(8):123
Shimomura, Tatsuo (EBARA Research Co Ltd/JAPAN) 5(2):129
Shinmura, Hiroshi (Ebara Corporation/JAPAN) 5(2):129
Shoemaker, Christine (Cornell University/USA) 5(2):27
Shooner, Frédéric (École Polytechnique de Montréal/CANADA) 5(5):135
Siboe, George M. (University of Nairobi/KENYA) 5(7):69
Siegel, Randy (URS Greiner Woodward Clyde/USA) 5(8):167
Siegrist, Robert (Colorado School of Mines/USA) 5(1):59
Sims, Ronald C. (Utah State University/USA) 5(5):97
Singletary, Michael A. (Duke Engineering & Services, Inc./USA) 5(1):89
Sinke, Anja J.C. (TNO Institute of Environmental Science/NETHERLANDS) 5(1):355
Skeen, Rodney S. (Battelle PNWD/USA) 5(2):165
Slack, William W. (FRx Inc./USA) 5(2):263
Sleeper, Pamela (U.S. Army Corps of Engineers/USA) 5(7):7
Slenders, Hans (TNO Institute of Environmental Sci/NETHERLANDS) 5(2):141
Smalley, J. Bryan (ERM-Southwest/USA) 5(1):349
Smets, Barth F. (University of Connecticut/USA) 5(7):27
Smit, A.L. (AB-DLO/NETHERLANDS) 5(3):529
Smith, David C. (University of Rhode Island/USA) 5(8):223

Smith, Gregory (ENSR Corporation/USA) 5(3):149
Smith, Steve (King's College/UK) 5(8):111, 123
Smith, Val H. (University of Kansas/USA) 5(1):283
Smolders, Eric (Catholic University Leuven/BELGIUM) 5(8):179
Snyder, Frances A. (Battelle/USA) 5(2):7
Solano-Serena, Floriane (Institut Français du Pétrole/FRANCE) 5(1):177
Song, Ji Hyeon (The University of Texas at Austin/USA) 5(5):117
Sorenson, Kent S. (Lockheed Martin Idaho Technologies Co./USA) 5(2):147
Sousa, Sofia (University of Aberdeen/UK) 5(3):369
Spain, Jim (U.S. Air Force/USA) 5(7):7
Spargo, Barry J. (U.S. Navy/USA) 5(8):223, 295
Sparrevik, Magnus (Norwegian Geotechnical Institute/NORWAY) 5(1):213
Spasova, Irena Ilieva (University of Mining & Geology/BULGARIA) 5(4):103
Speitel, Gerald E. (The University of Texas at Austin/USA) 5(7):217
Spexet, AnnMarie H. (Rice University/USA) 5(1):7
Spinnler, Gerard E. (Equilon Enterprises LLC [Shell/Texaco]/USA) 5(3):37
Spychala, Andrzej (Military University of Technology/POLAND) 5(5):1
Srinivasan, Usha (Yukon College/CANADA) 5(5):103
Stachowski, James R. (Acres International Corporation/USA) 5(3):65
Stadlbauer, Ernst A. (University of Applied Sciences/GERMANY) 5(4):97
Staley, James T. (University of Washington/USA) 5(8):191
Stamets, P. (Fungi Perfecti/USA) 5(6):63
Stanforth, Robert R. (National University of Singapore/SINGAPORE) 5(4):7
Staponski, Rod (EarthTech/USA) 5(7):63
Stearns, Steve M. (Equilon Enterprises LLC [Shell/Texaco]/USA) 5(3):37
Steele, Julia K. (U.S. Navy/USA) 5(8):223, 295
Steffan, Robert J. (Envirogen, Inc./USA) 5(6):1
Stegmann, Rainer (Technical University of Hamburg-Harburg/GERMANY) 5(5):25, 31, 43
Stehmeier, Lester G. (Nova Research & Technology Corp./CANADA) 5(1):207
Steinle, Patrick (University of Zurich/SWITZERLAND) 5(7):101
Stephen, John R. (University of Tennessee/USA) 5(3):345
Stieber, Michael (DVGW Technologiezentrum Wasser/GERMANY) 5(3):337
Stokes, J. (Lancaster University/UK) 5(8):247
Stokman, Gerard (SBNS/NETHERLANDS) 5(3):395
Strand, Stuart E. (University of Washington/USA) 5(6):133; 5(8):191
Strauss, Harald (University of Bochum/GERMANY) 5(3):475
Stright, Lisa (Michigan Technological University/USA) 5(3):97
Strobel, David A. (Mississippi State University/USA) 5(7):75
Stucki, Gerhard (CIBA Specialty Chemicals Inc./SWITZERLAND) 5(7):101
Studer, James E. (Duke Engineering & Services/USA) 5(1):89
Sucevich, Vaughn A. (EverClear Environmental Technologies/USA) 5(3):375
Suidan, Makram T. (University of Cincinnati/USA) 5(3):221; 5(8):283

Sullivan, Patricia E. (Pacific Gas and Electric Company/USA) 5(1):265
Sun, Yunwei (Battelle PNNL/USA) 5(1):29, 83
Sunderland, Garry (EA Technology Ltd./UK) 5(4):127
Sung, Kijune (Texas A&M University/ USA) 5(6):9
Surmacz-Górska, Joanna (Technical University of Silesia/POLAND) 5(5):1
Sveum, Per (Statskog Miljo og anlegg/ NORWAY) 5(8):99
Swadley, Lori A. (Colorado School of Mines/USA) 5(2):293
Swannell, Richard P.J. (AEA Technology/UK) 5(3):239, 251
Swanson, A.R. (University of Illinois/ USA) 5(2):21
Swingler, S.G. (The National Grid Co./UK) 5(8):259
Syslo, John (Roy F. Weston/REAC/ USA) 5(8):51
Syverson, Tim (GeoSyntec Consultants/ USA) 5(8):167
Szanislo, Paul J. (The University of Texas/USA) 5(5):123

Taat, Jan (Delft Geotechnics/ NETHERLANDS) 5(2):73
Tabak, Henry H. (U.S. EPA/USA) 5(4):37; 5(8):283
Tacy, Peter D. (STS Consultants, Ltd./ USA) 5(2):15
Taghon, Gary L. (Rutgers, The State University of New Jersey/USA) 5(8):271
Talley, Jeffrey W. (U.S. Army Corps of Engineers/USA) 5(3):545; 5(5):87; 5(8):289
Tamburini, Davide (Battelle Europe/ SWITZERLAND) 5(1):159
Tan, H-M. (National University of Singapore/SINGAPORE) 5(3):389
Tan, Nico C.G. (Wageningen Agricultural University/ NETHERLANDS) 5(7):253
Tan, Yoke-Cheng (DSO National Laboratories/SINGAPORE) 5(3):257
Tang, Jinshan (Hydro Geo Chem, Inc./ USA) 5(3):59
Tartakovsky, Boris (Biotechnology Research Institute/CANADA) 5(7):107
te Stroet, C.B.M. (TNO Institute of Applied Geoscience/ NETHERLANDS) 5(2):141
Tetreault, A. (École Polytechnique de Montréal/CANADA) 5(3):451
Thalmann, Philipp (Swiss Federal Institute of Technology/ SWITZERLAND) 5(7):101
Tharakan, John P. (Howard University/ USA) 5(7):179
Thiboutot, Sonia (National Defense/ CANADA) 5(7):15
Thomas, Paul R. (Thomas Consultants, Inc./USA) 5(6):21, 115
Thomas, RueAnn (International Paper/ USA) 5(7):83; 5(8):167
Thomas, Susan (Battelle/USA) 5(6):63
Thompson, Ian D. (Morrow Environmental Consultants Inc./ CANADA) 5(1):147
Thompson, Leslie C. (Pintail Systems, Inc./USA) 5(4):31
Thompson, Scott E. (Malcolm Pirnie, Inc./USA) 5(7):7
Thornton, Crisandel F. (Clark Atlanta University/USA) 5(4):109
Thornton, Steven F. (University of Sheffield/UK) 5(1):277
Thurman, E. Mike (U.S. Geological Survey/USA) 5(1):283
Ting, Yen-Ping (National University of Singapore/SINGAPORE) 5(3):389
Tonnaer, Haimo (TAUW Infra Consult bv/NETHERLANDS) 5(1):355; 5(3):445
Torbeyns, Bert (Catholic University Leuven/BELGIUM) 5(8):179
Torres, Luis Gilberto (Instituto Mexicano del Petroleo/MEXICO) 5(7):137; 5(8):197
Torres, Michael (U.S. EPA/USA) 5(6):89
Townsend, Richard Todd (Texas A&M University/USA) 5(3):209; 5(6):57

Trepte, Björn (Stockholm University/SWEDEN) 5(8):229
Trovato, Antonino (Battelle Europe/ SWITZERLAND) 5(5):69
Trudnowski, John (MSE Technology Applications, Inc./USA) 5(6):205
Truong, Tri Vu (Sodexen Group/ CANADA) 5(8):1
Tsuji, Hirokazu (Obayashi Corporation/ JAPAN) 5(5):13
Tucker, Jason L. (University of Georgia/ USA) 5(6):145
Tuovinen, Olli (The Ohio State University/USA) 5(6):181
Turek-Szytow, Jolanta (Silesian Technical University/POLAND) 5(8):203
Tustin, John (Office of Worcester County Commisioners/USA) 5(6):251

Ünlü, Kahraman (Middle East Technical University/TURKEY) 5(5):81
Unterman, Ronald (Envirogen, Inc./ USA) 5(7):7
Unz, Richard (The Pennsylvania State University/USA) 5(4):147

Vaa, Arne (Vaa Biomiljo AS/ NORWAY) 5(5):37
Valocchi, Albert J. (University of Illinois at Urbana-Champaign/USA) 5(1):301
van Bavel, Bert (Umea University/ SWEDEN) 5(8):135
Vandecasteele, Jean-Paul (Institut Français du Pétrole/FRANCE) 5(1):177; 5(3):25; 5(8):117
van den Toorn, Antonie (SC-DLO/ NETHERLANDS) 5(6):39; 5(8):87
Vanderby, Philippe (École Polytechnique de Montréal/ CANADA) 5(3):177
van der Grift, Bas (TNO Institute of Applied Geoscience/ NETHERLANDS) 5(3):463
van der Hoek, Eline E. (KEMA/ NETHERLANDS) 5(3):395
van der Kooij, Aldert (DHV Environment and Infrastructure/ NETHERLANDS) 5(8):87
van der Waarde, Jaap J. (Bioclear Environmental Biotechnology/ NETH-ERLANDS) 5(2):225; 5(3):395; 5(6):39
van der Werf, A.W. (Bioclear Environmental Biotechnology/ NETHERLANDS) 5(2):225
Van Geel, Paul J. (Carleton University/ CANADA) 5(1):77, 325
van Gestel, Kees (Vrije Universiteit van Amsterdam/NETHERLANDS) 5(3):395
van Heiningen, Erwin (TNO Institute of Environmental Sci/NETHERLANDS) 5(2):73; 5(3):481
van Leeuwen, J. (Hans) (University of New England (Australia)/ AUSTRALIA) 5(5):203
van Liere, Herco C. (TNO Institute of Environmental Sci/NETHERLANDS) 5(6):81
van Lieshout, Johan (Wageningen Agricultural University/ NETHERLANDS) 5(8):69
van Ree, C.C.D.F. (Derk) (Delft Geotechnics/NETHERLANDS) 5(3):83, 103, 109
van Veen, H. Johan (TNO Institute of Environmental Science/ NETHERLANDS) 5(1):355
van Winden, Bas (Arcadis/ NETHERLANDS) 5(2):73
van Zyl, Heleen D. (Technikon Pretoria/ REP. OF SOUTH AFRICA) 5(5):51
Vardy, James A. (U.S. Coast Guard Support Center/USA) 5(1):115
Vázquez-Duhalt, Rafael (Universidad Nacional Autonoma de Mexico/ MEXICO) 5(3):409; 5(5):211
Veltkamp, A.G. (Nederlandse Aardolie Maatschappij BV/NETHERLANDS) 5(3):481; 5(6):231
Venosa, Albert (U.S. EPA/USA) 5(3):221
Verheul, Jos H.A.M. (NOBIS/ NETHERLANDS) 5(6):295
Vermeulen, Bert (IMAG-DLO/ NETHERLANDS) 5(6):39
Vermeulen, H.J. (NOBIS/ NETHERLANDS) 5(6):295

Vesper, Stephen J. (U.S. EPA/USA) 5(2):263
Vethe, Øistein (Jordsforsk/NORWAY) 5(8):99
Vetter, Yves-Alain (University of Washington/USA) 5(8):301
Veul, Martin F. X.W. (Witteveen & Bos Consulting Engineers/ NETHERLANDS) 5(3):395
Viebranz, N. (Umweltschutz Nord GmbH & Co/GERMANY) 5(5):43
Villani, Marco (Centro Ricerche Ambientali Montecatini/ITALY) 5(3):511
Villemur, Richard (INRS-Institut Armand-Frappier/CANADA) 5(2):243; 5(7):89
Vincent, Jennifer C. (Earth Tech/USA) 5(2):255
Vogt, Gerald T. (CH2M Hill/USA) 5(4):91
Vreeken, C. (BMS BodemMilieuSystemen/ NETHERLANDS) 5(3):103

Wakeman, John S. (U.S. Army Corps of Engineers/USA) 5(8):315
Wallace, Mark N. (U.S. Army Corps of Engineers/USA) 5(3):505
Wallace, S. (Komex Clark Bond/UK) 5(8):123
Walsh, Jami B. (Worcester Polytechnic Institute/USA) 5(3):265
Walter, David J. (C-CORE Research Institute/CANADA) 5(2):249
Walter, Gary R. (Hydro Geo Chem, Inc./USA) 5(3):59
Walworth, James L. (University of Arizona/USA) 5(3):505, 523
Wang, Chuhua (University of Georgia/ USA) 5(6):239
Wang, M.C. (The Pennsylvania State University/USA) 5(2):237
Wang, Sookyun (Texas A&M University/USA) 5(2):269
Wang, Xiaoping (University of Washington/USA) 5(6):133
Ward, Dawn M. (University of Delaware/USA) 5(8):223
Wardwell, David (Mission Research Corporation/USA) 5(3):47
Wasson, Louis (Mississippi State University/USA) 5(7):75
Watson, George (U.S. Navy/USA) 5(3):201
Watson, Jonathan S. (University of Newcastle-upon-Tyne/UK) 5(3):251
Watts, Richard J. (Washington State University/USA) 5(7):211
Wayment, Darcey G. (Rice University/ USA) 5(6):121
Weber, Bernd (University of Applied Sciences/GERMANY) 5(4):97
Weber, Walter J. (University of Michigan/USA) 5(6):75
Weigner, Pavel (Institute of Chemical Technology/CZECH REPUBLIC) 5(5):161
Weitz, Hedda J. (University of Aberdeen/UK) 5(3):369; 5(8):217
Werner, Peter (Technische Universität Dresden/GERMANY) 5(3):337; 5(8):117
Wharry, Stan (U.S. Army/USA) 5(7):63
Wheater, H.S. (Imperial College of Sci., Tech., and Med./UK) 5(8):123
White, Daniel M. (University of Alaska/ USA) 5(3):499, 523
White, David C. (University of Tennessee/USA) 5(3):345
Wickramanayake, Godage B. (Battelle/USA) 5(3):89
Widdowson, Mark A. (Virginia Tech/USA) 5(6):45; 5(8):19
Wieck, James M. (GZA GeoEnvironmental, Inc./USA) 5(2):171
Wiedemeier, Todd H. (Parsons Engineering Science, Inc./USA) 5(1):121, 313
Wildeman, Thomas (Colorado School of Mines/USA) 5(6):217
Williamson, Kenneth J. (Oregon State University/USA) 5(7):131
Willis, Matthew B. (Cornell University/ USA) 5(2):27
Willis, Michael (MSE Technology Applications, Inc./USA) 5(6):205

Wilson, Barbara H. (Dynamac Corporation/USA) 5(1):115
Wilson, David J. (Eckenfelder/Brown and Caldwell/USA) 5(1):53, 59
Wilson, John T. (U.S. EPA/USA) 5(1):103; 5(1):109
Winningham, Jason (EnSafe, Inc./USA) 5(8):37
Winsor, L. (Memorial University of Newfoundland/CANADA) 5(1):207
Wisotzky, Frank (University of Bochum/GERMANY) 5(3):475
Woertz, Jennifer R. (The University of Texas at Austin/USA) 5(5):123
Wolf, Duane C. (University of Arkansas/USA) 5(6):33
Wolf, Joseph L. (Philips Components/USA) 5(1):13
Wolfe, N. Lee (U.S. EPA/USA) 5(6):145, 151, 239
Wong, Meng Meng (The University of Tennessee/USA) 5(7):161
Wong, Tai T. (O'Connor Associates Environmental Inc./CANADA) 5(5):111
Woodhull, Patrick M. (IT Corporation/USA) 5(7):33, 63
Woodward, David S. (Earth Tech/USA) 5(3):143
Woolard, Craig (University of Alaska/USA) 5(3):499, 505, 523
Word, Jack (Battelle/USA) 5(6):63
Wrenn, Brian A. (Washington University/USA) 5(3):221
Wright, Christopher F. (GZA GeoEnvironmental, Inc./USA) 5(2):171
Wu, Jia-Ling (Jacobs Engineering/USA) 5(1):153
Wu, Lin (University of California Davis/USA) 5(4):141
Wu, Madeline (Water Restoration, Inc./USA) 5(2):177

Xing, Jian (Global Remediation Technologies, Inc./USA) 5(2):217
Yagi, Osami (National Institute for Environmental Studies & CREST/JAPAN) 5(2):89, 95, 101; 5(3):233
Yamane, Hisakazu (The University of Tokyo/JAPAN) 5(8):155
Yang, Chung-Chih (National Chung Hsing University/TAIWAN-R.O.C.) 5(2):205
Yang, Wei (University of Cincinnati/USA) 5(4):37
Yare, Bruce S. (Solutia Inc./USA) 5(5):19
Yeh, Mon-Fan (National Chung Hsing University/TAIWAN - R.O.C.) 5(2):205
Yen, Teh Fu (University of Southern California/USA) 5(6):175
Yerushalmi, Laleh (Biotechnology Research Institute/CANADA) 5(3):439
Yilmaz, Ayse Hilal (Ege University/TURKEY) 5(5):149
Young, Lily Y. (Rutgers University/USA) 5(8):271
Zagury, Gérald J. (École Polytechnique de Montréal/CANADA) 5(6):211
Zaidel, J. (Waterloo Hydrogeologic, Inc./USA) 5(2):281
Zaluski, Marek (MSE Technology Applications/USA) 5(6):205
Zamojski, Lawrence D. (Acres International Corporation/USA) 5(3):65
Zapf-Gilje, Reidar (Golder Associates Ltd. & University of British Columbia/CANADA) 5(1):231
Zare, Richard N. (Stanford University/USA) 5(8):289
Zawislanski, Peter T. (Lawrence Berkeley National Laboratory/USA) 5(4):141
Zenker, Matthew J. (North Carolina State University/USA) 5(1):165
Zhang, Tian C. (University of Nebraska - Lincoln/USA) 5(4):85
Zubrow, Jim (Key Environmental Inc./USA) 5(1):289
Zweerts, Herm (Catholic University Leuven/BELGIUM) 5(8):179
Zwick, Thomas C. (Battelle/USA) 5(3):89, 201; 5(5):57
Zwiren, William N. (CH2M HILL/USA) 5(3):77

KEYWORD INDEX

This index contains keyword terms assigned to the articles in the eight-volume proceedings of the Fifth International In Situ and On-Site Bioremediation Symposium (San Diego, California, April 19-22, 1999). Ordering information is provided on the back cover of this book.

In assigning the terms that appear in this index, no attempt was made to reference all subjects addressed. Instead, terms were assigned to each article to reflect the primary topics covered by that article. Authors' suggestions were taken into consideration and expanded or revised as necessary. The citations reference the eight volumes as follows:

5(1): Alleman, B.C., and A. Leeson (Eds.), *Natural Attenuation of Chlorinated Solvents, Petroleum Hydrocarbons, and Other Organic Compounds*. Battelle Press, Columbus, OH, 1999. 402 pp.

5(2): Leeson, A., and B.C. Alleman (Eds.), *Engineered Approaches for In Situ Bioremediation of Chlorinated Solvent Contamination*. Battelle Press, Columbus, OH, 1999. 336 pp.

5(3): Alleman, B.C., and A. Leeson (Eds.), *In Situ Bioremediation of Petroleum Hydrocarbon and Other Organic Compounds*. Battelle Press, Columbus, OH, 1999. 588 pp.

5(4): Leeson, A., and B.C. Alleman (Eds.), *Bioremediation of Metals and Inorganic Compounds*. Battelle Press, Columbus, OH, 1999. 190 pp.

5(5): Alleman, B.C., and A. Leeson (Eds.), *Bioreactor and Ex Situ Biological Treatment Technologies*. Battelle Press, Columbus, OH, 1999. 256 pp.

5(6): Leeson, A., and B.C. Alleman (Eds.), *Phytoremediation and Innovative Strategies for Specialized Remedial Applications*. Battelle Press, Columbus, OH, 1999. 340 pp.

5(7): Alleman, B.C., and A. Leeson (Eds.), *Bioremediation of Nitroaromatic and Haloaromatic Compounds*. Battelle Press, Columbus, OH, 1999. 302 pp.

5(8): Leeson, A., and B.C. Alleman (Eds.), *Bioremediation Technologies for Polycyclic Aromatic Hydrocarbon Compounds*. Battelle Press, Columbus, OH, 1999. 358 pp.

A

A. bisporus, see Agaricus bisporus
accumulation ***5(5):***117
acid mine drainage, *see* mine waste
acidophiles ***5(6):***225
activation ***5(7):***89
adsorption ***5(8):***271
adsorption/desorption equilibria and kinetics ***5(8):***283
aeration ***5(3):***103
aerobic ***5(2):***81; ***5(7):***27; ***5(8):***197
aerobic bioremediation ***5(3):***295
aerobic degradation ***5(6):***265
aerobic stabilization ***5(6):***245
Agaricus bisporus ***5(8):***87
aging ***5(8):***235
air channeling ***5(3):***103
air sparging ***5(3):***47, 59, 83, 89, 103, 109, 123, 169, 445, 541
air sparging/soil vapor extraction pilot study (AS/SVE pilot study) ***5(3):***307

alachlor ***5(1):***283
alkanes ***5(3):***7
alternative electron acceptor ***5(3):***475
aluminosilicate ***5(4):***79
ammonium nitrate ***5(4):***91
anaerobic bioreactor ***5(6):***217
anaerobic oxidation ***5(2):***73
anaerobic/aerobic ***5(7):***45, 253
anaerobic/aerobic composting ***5(7):***125
anaerobic/aerobic in situ treatment ***5(6):***1
analytical model ***5(1):***83
anoxic ***5(7):***27
anthracene oil ***5(8):***203
arctic ***5(3):***227, 493 (*see also* cold temperature)
Arochlor™ ***5(7):***179
aromatic(s) ***5(3):***357, 535
arsenic ***5(5):***7
Arthrobacter sp. ***5(7):***75
artificial neural network ***5(4):***1
AS/SVE pilot study, *see* air sparging/soil vapor extraction pilot study
assessment ***5(8):***253
attenuation ***5(1):***97, 147; ***5(8):***223
automation control system ***5(3):***177
autotrophic denitrification ***5(4):***85
azo dye ***5(7):***253
Azospirillum ***5(3):***331

B

B. cepacia, see Burkholderia cepacia
BAC, *see* biological activated carbon
bacterial transport ***5(2):***231
barium ***5(3):***383
barrier ***5(6):***169
batch reactors ***5(7):***217
beaches ***5(3):***239
bedrock ***5(1):***71
BEHP, *see bis*(2-ethylhexyl)phthalate
bench-scale column ***5(3):***195
benzene ***5(1):*** 97, 207, 253, 265, 271; ***5(3):***535; ***5(6):***81
benzene biodegradation ***5(3):***481
benzene-to-toluene ratio ***5(1):***141
benzene, toluene, ethylbenzene, and xylenes (BTEX) ***5(1):***135, 183, 189, 195, 201, 213, 231, 325; ***5(3):***77, 89, 183, 265, 277, 283, 289, 295, 319, 337, 363, 369, 375, 427, 451, 463, 475; ***5(5):***161; ***5(8):***141
benzo(a)pyrene ***5(8):***93, 179
binding ***5(8):***75
bioactivation ***5(6):***211
bioassay ***5(3):***395; ***5(8):***57
bioattenuation ***5(1):***201
bioaugmentation ***5(2):***211, 231, 269, 293; ***5(3):***409, 415, 421, 433; ***5(4):***13; ***5(6):***139; ***5(7):***101, 119, 173
bioavailability ***5(1):***307; ***5(3):***313; ***5(4):***25; ***5(7):***155, 229; 241, 247, 259; ***5(8):***57, 69, 87, 161, 217, 241, 247, 253, 259, 265, 271, 283
bioavailable iron ***5(1):***331
biobarrier(s) ***5(2):***211, 243, 269; ***5(3):***37, 439, 445; ***5(4):***13, 47, 53, 109; ***5(6):***175, 37
BIOCHLOR ***5(1):***83
biodegradation rate ***5(3):***301
biodiversity***5(2):***243
biofilm(s) ***5(2):***243; ***5(7):***27
biofilm reactor ***5(6):***225
biofiltration ***5(3):***451; ***5(5):***111, 117, 123, 135, 203; ***5(7):***137
biogas ***5(5):***149
biogenic interference ***5(3):***523
bioleaching ***5(4):***127
biological activated carbon (BAC) ***5(5):***203
bioluminescence, *see lux*-marked microorganisms
biomarkers ***5(1):***171
biomass ***5(5):***117; ***5(6):***39
biomass reduction ***5(6):***277
biomolecule ***5(2):***41
biopile ***5(5):***7, 43, 51, 57, 63, 69, 183; ***5(7):***33; ***5(8):***63
biopolymer ***5(6):***175
bioreactor ***5(4):***147; ***5(5):***123, 143; ***5(6):***181, 205, 265, 277; ***5(7):***107, 113, 253
biorecovery ***5(4):***37
bioredox ***5(1):***77
bioregeneration ***5(7):***217
bioreporter ***5(8):***277
BIOSCREEN ***5(1):***115
biosensor ***5(3):***351, 369; ***5(4):***25; ***5(7):***247

bioslurping *5(3):*131, 137, 143, 155, 277
bioslurry reactor *5(5):*19, 25; *5(7):*15, *5(8):*7, 135
biosparging *5(2):*217, 255; *5(3):*97, 109, 115, 541; *5(8):*129, 167
biosurfactant *5(3):*457
biotrickling filter *5(5):*129
bioventing *5(2):*7, 281; *5(3):*115, 137, 149, 169, 177, 183, 189, 195, 201, 383, 529, 541; *5(5):*183; *5(6):*245, *5(8):*167
bioventing efficiency *5(3):*195
biphenyl *5(7):*179
bis(2-ethylhexyl)phthalate (BEHP) *5(5):*19
Brazilian gasoline *5(1):*135
BTEX, *see* benzene, toluene, ethylbenzene, and xylenes
bunker fuel *5(3):*189, 201
Burkholderia cepacia *5(2):*293
n-butanol *5(2):*89

C

cadmium (Cd) *5(4):*103
calcium peroxide *5(7):*203
carbon tetrachloride *5(2):*35, 41; *5(6):*133
carboxylic acids *5(3):*251
catalyst *5(4):*97
Cd, *see* cadmium
cellular automata *5(3):*511
characterization *5(8):*241
chelate *5(8):*229
Chem-Bio *5(8):*147
chemical countermeasures *5(6):*57
chemical oxidation *5(2):*199
chemical oxygen demand (COD) *5(5):*203
chlorate *5(4):*147
chlorinated ethenes *5(2):*165
chlorinated hydrocarbons *5(2):*1, 61, 255
chlorinated phenols *5(7):*95
chlorinated solvents *5(1):*19, 29, 53, 59, 65, 71; *5(2):*121, 171, 185, 281, 287; *5(3):*541
chlorinated VOC(s) (CVOC) *5(1):*247; *5(2):*135
o-chloroaniline *5(7):*223
chlorobenzene *5(6):*109; *5(7):*223
chloroform *5(2):*41
chlorophenol(s) *5(7):*101, 113, 191
2-chlorophenol *5(7):*223
chromated copper arsenate *5(4):*121
chromium reduction *5(4):*13
chromium(VI) *5(4):*121, 135
cis-1,2-dichloroethene (*cis*-DCE) *5(2):*81, 181, 185, 225
coal tar *5(1):*289
coastal *5(3):*221
COD, *see* chemical oxygen demand
cold temperature *5(1):*195; *5(3):*499, 505; *5(5):*63; *5(6):*187
colloid *5(2):*231
combined technology *5(7):*211
cometabolism *5(1):*89, 253; *5(2):*1, 113, 217, 281; *5(3):*7, 31, 541; *5(7):*131, 179
commingled plumes *5(1):*247
community structure *5(6):*231
composting *5(5):*13, 31, 37, 189; *5(7):*33, 51, 57, 63; *5(8):*25, 31
computer model *5(2):*281
concrete decontamination *5(7):*143
consortium/consortia *5(3):*245
constant flow *5(3):*195
containment *5(7):*167
contaminant transport *5(8):*283
contingency plan *5(1):*313
copper (Cu) *5(4):*25, 103
cosolvency *5(1):*135
cost-benefit analysis *5(6):*289
cosubstrate *5(1):*165
cover *5(5):*63
Cr(VI), *see* chromium(VI)
creosote *5(1):*289; *5(6):*45; *5(7):*75; *5(8):*37, 99
crop plants *5(6):*15
crude oil *5(1):*159, 359; *5(3):*209, 221, 233, 239, 245, 251, 351, 421; *5(5):*69, 189; *5(6):*57; *5(8):*51
Cu, *see* copper
CVOC, *see* chlorinated VOC(s)
cyanide *5(4):*31
cycled air sparging *5(3):*77
cyclodextrins *5(8):*247

D

D. frappieri, see Desulfitobacterium frappieri

2,4-D, *see* 2,4-dichlorophenoxyacetic acid
1,1-DCE, *see* 1,1-dichloroethene
1,2-DCA, *see* 1,2-dichloroethane
1,2-DCE, *see* 1,2-dichloroethene
DCP, *see* dichlorophenol
DDD, *see* dichlorodiphenyldichloroethane
DDE, *see* 1,1-dichloro-2,2-*bis*(*p*-chlorophenyl)ethylene
DDT, *see* dichlorodiphenyltrichloroethane
dechlorination ***5(2):***41; ***5(7):***83, 185, 259
dehalogenase ***5(6):***145
dehalogenation ***5(2):***171
dehydrogenase ***5(3):***535
demeton ***5(6):***151
denitrification ***5(2):***73; ***5(3):***325; ***5(4):***53, 59, 65, 73, 79, 91
denitrifier ***5(8):***191
dense, nonaqueous-phase liquid (DNAPL) ***5(1):***295; ***5(2):***55, 191
desorption ***5(7):***259; ***5(8):***265
Desulfitobacterium frappieri ***5(7):***107
dichlorobenzene ***5(2):***7; ***5(5):***135
1,1-dichloro-2,2-*bis*(*p*-chlorophenyl)ethylene (DDE) ***5(7):***69
dichlorodiphenyldichloroethane (DDD) ***5(7):***69
dichlorodiphenyltrichloroethane (DDT) ***5(7):***69, 125, 137
1,2-dichloroethane (1,2-DCA) ***5(2):***73; ***5(6):***109; ***5(7):***223
1,1-dichloroethene and 1,1,-dichloroethylene (1,1-DCE) ***5(1):***71
1,2-dichloroethene and 1,2,-dichloroethylene (1,2-DCE) ***5(1):***65
dichlorophenol (DCP) ***5(1):***19; ***5(7):***83
2,4-dichlorophenol (2,4-DCP) ***5(7):***247
2,4-dichlorophenoxyacetic acid (2,4-D) ***5(6):***265; ***5(7):***119, 137
1,2-dichloropropane ***5(1):***19; ***5(7):***223
diesel ***5(3):***409, 457; ***5(5):***51
diffusion ***5(8):***283
dihaloelimination ***5(2):***73
2,4-dimethylphenol ***5(8):***13
dinitrotoluene (DNT) ***5(7):***7, 27, 33
1,4-dioxane ***5(1):***165; ***5(6):***139
dioxygenase ***5(3):***357; 535
direct push ***5(1):***337
dissolved hydrogen ***5(1):***331
diversity index ***5(3):***233
DNA fingerprinting ***5(5):***143
DNAPL, *see* dense, nonaqueous-phase liquid
DNT, *see* dinitrotoluene
dodecane ***5(3):***389
drinking water ***5(4):***59
dry cleaners ***5(1):***7
dual permeability ***5(3):***103
dual-phase extraction ***5(3):***137, 155
dye decolorization ***5(5):***211
dyes ***5(6):***69, 271
dynamic system ***5(3):***177
dynamics model ***5(2):***287

E

E. canadensis, see Elodea canadensis
ecological risk ***5(3):***395
economics ***5(6):***289
ecotoxicity ***5(2):***243
EDTA, *see* ethylenediaminetetraacetic acid
electrochemical ***5(3):***375
electrode ***5(4):***73
electrokinesis ***5(4):***127
electron acceptors ***5(1):***189
electron donors ***5(1):***59; ***5(2):***47; ***5(4):***135
electron paramagnetic resonance (EPR) spectroscopy (EPR) ***5(8):***75
electrophoretic mobility ***5(6):***163
Elodea canadensis ***5(6):***145
endpoints ***5(7):***185
enhanced anaerobic dechlorination ***5(2):***135
enrichment ***5(3):***245; ***5(7):***149
environmental estrogens ***5(7):***155
environmental impact ***5(3):***541
EPA protocol ***5(1):***53
EPR, *see* electron paramagnetic resonance spectroscopy
ergosterol ***5(5):***103
ETBE, *see* ethyl *tert*-butyl ether
ethane-utilizing bacterium ***5(2):***89
ethanol ***5(1):***135
ethers ***5(1):***165; ***5(7):***131
ethylene ***5(2):***225

ethylenediaminetetraacetic acid (EDTA) ***5(8):***229
ethyl *tert*-butyl ether (ETBE) ***5(3):***25
evapotranspiration ***5(6):***115
explosives ***5(6):***121; ***5(7):***1, 21, 33, 39, 45, 51, 57, 63, 197, 203
extraction aging ***5(8):***253

F

FAME, *see* fatty acid methyl ester
fate ***5(6):***109
fatty acid methyl ester (FAME) profile(s) ***5(6):***33
Fe(III), *see* iron(III)
Fenton's reagent ***5(2):***191; ***5(6):***157; ***5(7):***197, 211
fermentation ***5(5):***149
field scale, field-scale ***5(1):***213
field studies ***5(1):***7; ***5(6):***57
filtration ***5(2):***231
flow sensor ***5(3):***47
fluid injection ***5(2):***249
fluidized-bed bioreactor ***5(7):***7
fluoranthene ***5(8):***179
fluorescent primers ***5(5):***143
fractured media ***5(1):***71, 183; ***5(2):***255
free-product recovery ***5(3):***131
fuel ***5(5):***129
fuel hydrocarbons ***5(1):***121
fuel oil ***5(3):*** 115, 227, 257, 493
full scale, full-scale ***5(2):***113
full-scale air sparging/soil vapor extraction (AS/SVE) ***5(3):***307
full-scale AS/SVE, *see* full-scale air sparging/soil vapor extraction
fungus (fungi) ***5(1):***147; ***5(5):***103, 123, 203, 211; ***5(6):***63, 69; ***5(7):***173; ***5(8):***75, 105
funnel-and-gate ***5(4):***109

G

GAC, *see* granular activated carbon
gaseous nutrient injection ***5(2):***107
gasoline ***5(1):***177; ***5(5):***111
gasoline additive ***5(3):***25
gas-phase, *see* vapor-phase
GEM, *see* genetically engineered microorganism
genetic algorithms ***5(2):***275
genetically engineered microorganism (GEM) ***5(8):***277
geochemistry ***5(1):***59
granular activated carbon (GAC) ***5(7):***217
granular peat moss ***5(3):***451; ***5(6):***211
grasses ***5(6):***45
groundwater ***5(1):***13, 41, 271, 295

H

H_2O_2, *see* hydrogen peroxide
half-life ***5(1):***97, 259
halo-elimination ***5(1):***89
HCA™ ***5(6):***145
HCH, *see* hexachlorocyclohexane
heap treatment ***5(5):***7
heat ***5(5):***63
heat transfer ***5(1):***201
heating oil ***5(3):***271
heavy metals ***5(3):***389; ***5(4):***13, 19, 25, 121; ***5(5):***197; ***5(6):***211
herbicides ***5(1):***283; ***5(7):***143
heterogeneity ***5(1):***47
hexachlorobenzene ***5(7):***259
hexachlorocyclohexane (HCH) ***5(6):***81
high explosives ***5(7):***217
HMX ***5(7):***1, 15, 21, 45, 217
Hormoconis ***5(1):***147
HRC, *see* hydrogen-release compound
humic acid ***5(8):***161
humification ***5(7):***45
hybrid poplar ***5(6):***21, 115, 127
hybrid willow ***5(6):***21
hydraulic fracture ***5(2):***263
hydrocarbon ***5(1):***109; ***5(8):***203
hydrogen ***5(2):***185; ***5(4):***147
hydrogen peroxide ***5(3):***331, 337
hydrogen-release compound (HRC) ***5(2):***61, 67, 121, 177, 181, 287
hydroxyl radical ***5(3):***375

I

IAS, *see* air sparging
immobilization ***5(5):***161, 197
in situ air sparging, *see* air sparging
induction ***5(8):***25
intermittent flow ***5(3):***195

intrinsic bioremediation ***5(1):***65, 231, 301; ***5(2):***73; ***5(3):***529; ***5(8):***19, 223
iron(III) [Fe(III)] ***5(1):***307
iron oxidation ***5(8):***141
iron-reducing ***5(1):***189
isotope(s) ***5(5):***189

J

jet fuel (*see also* JP-4, JP-5) ***5(1):***147
JP-4 ***5(3):***183, ***5(3):***345
JP-5 ***5(3):***383, ***5(3):***131

K

kerosene ***5(3):***529, ***5(3):***195
kinetics ***5(2):***21; ***5(3):***97, 245, 463; ***5(4):***65; ***5(7):***229; ***5(8):***173, 301

L

laboratory-scale ***5(3):***403
laccase(s) ***5(5):***211; ***5(8):***105
lactate ***5(2):***165; ***5(4):***135
land treatment ***5(3):***215; ***5(5):***81, 97, 189
land use ***5(6):***39
landfarming ***5(5):***13, 75, 87, 177; ***5(6):***39; ***5(8):***37, 43, 57
landfill ***5(6):***169, 245
landfill leachate ***5(6):***251; 259
Lasagna™ ***5(2):***263
LCA, *see* life-cycle assessment(s) ***5(3):***541
LC/ESI/MS, *see* liquid chromatography/electrospray ionization/mass spectrometry
leaching ***5(4):***103
lead (Pb) ***5(4):***7, 103; ***5(5):***7; ***5(6):***217
life-cycle assessment(s) (LCA) ***5(3):***541
light nonaqueous-phase liquid (LNAPL) ***5(3):***143, 149
Lignicolous ***5(7):***69
lignin peroxidase ***5(8):***105
liquefaction ***5(6):***175
liquid chromatography/electrospray ionization/mass spectrometry (LC/ESI/MS) ***5(1):***171
LNAPL, *see* light nonaqueous-phase liquid
LNAPL recovery ***5(3):***163
long-term monitoring ***5(1):***313
low temperature, *see* cold temperature
lux-marked microorganisms ***5(8):***217, 277

M

macrofauna ***5(8):***259
magnesium peroxide ***5(3):*** 1, 301
malathion ***5(6):***151
manganese oxidation ***5(8):***141
manganese peroxidase ***5(8):***105
manufactured gas plant (MGP) ***5(1):***265, 295; ***5(8):***117, 129, 147, 229
marine ***5(3):***227, 257, 493
marine biodegradation ***5(8):***301
marine sediment ***5(8):***301, 315
marsh ***5(3):***325
mass transfer ***5(3):***123
mathematical modeling ***5(2):***269
membrane ***5(6):***265
mesocosms ***5(1):***283
metabolites ***5(8):***135
metal mine drainage, *see* mine waste
metals ***5(4):***31, 109, 141; ***5(5):***7
methane ***5(2):***81, 113; ***5(3):***541
methane production ***5(5):***149
methanogenic ***5(2):***35, 41
methanol ***5(2):***157; ***5(7):***235
methanotroph(s) ***5(1):***253; ***5(2)***101, 107, 191, 255, 263
methanotrophic bacteria, *see* methanotroph(s)
methyl *tert*-butyl ether (MTBE) ***5(1):***97, 103, 109, 115, 121,165, 183; ***5(3):***1, 7, 13, 19, 31, 37, 89, 289, 427; ***5(5):***129
methylene chloride ***5(1):***13, 35
Methylocystis ***5(2):***101
metolachlor ***5(1):***283
MGP, *see* manufactured gas plant
microaggregation ***5(8):***179
microbial activity ***5(1):***47
microbial diversity ***5(6):***33; ***5(8):***209
microbial ecology ***5(3):***403
microbial mat ***5(4):***109
microbial transport ***5(2):***211
microcosm study ***5(2):***55; ***5(6):***109; ***5(8):***19
microflora ***5(8):***259
microscale analysis ***5(8):***289

Microtox™ ***5(4):***7; ***5(5):***37
mine waste ***5(4):***19, 31, 37; ***5(6):***193, 205, 211, 217, 225; ***5(7):***235
mineral oil ***5(3):***313, 395
mineralization ***5(8):***25, 123, 247, 295
mixed contaminants ***5(5):***1
mobility ***5(8):***235
modeling ***5(1):*** 13, 29, 115, 201, 259, 289, 301; ***5(2):***275; ***5(3):***59, 97, 511; ***5(4):***65, ***5(6):***9; ***5(7):***27, 185, 229, 259
moisture content ***5(5):***155
molecular monitoring ***5(3):***363
monitored natural attenuation, *see* natural attenuation
monitoring ***5(1):***177, 331, 343; ***5(3):***189, 357, 529; ***5(5):***103, 143; ***5(8):***209
monitoring design ***5(1):***301
monitoring well ***5(1):***337
monochlorobenzene ***5(6):***81
Monod ***5(8):***173
MTBE, *see* methyl *tert*-butyl ether
multiphase extraction ***5(3):***131, 277
Mycobacterium ***5(2):***89, 95
mycorrhiza ***5(6):***27

N

n-butanol, *see* butanol
nanofiltration ***5(4):***97
naphthalene ***5(7):***223; ***5(8):***191, 277
NAPL, *see* nonaqueous-phase liquid
natural attemation ***5(1):***13, 29, 35, 41, 47, 53, 59, 65, 77, 109, 115, 121, 141, 153, 159, 177, 207, 219, 225, 241, 247, 253, 259, 265, 271, 277, 283, 289, 295, 313, 325, 331, 355; ***5(6):***81; ***5(7):***83; ***5(8):***13
natural attenuation modeling ***5(1):***83
natural bioremediation ***5(3):***295
natural organic material ***5(3):***523
nematodes ***5(6):***231
neutron probe ***5(3):***169
nitrate (NO_3) ***5(3):***337, 469; ***5(4):***31, 47, 53, 59, 65, 97
nitrification ***5(7):***131
nitrilotriacetate (NTA) ***5(8):***229
2-nitroaniline ***5(7):***223
nitroaromatic ***5(7):***211, 223
nitrobenzene ***5(7):***95, 223
nitrogen ***5(3):***499
nitrotoluene ***5(7):***7
NO_3, *see* nitrate
nonaqueous-phase liquid (NAPL) ***5(8):***259
nonindigenous microorganisms ***5(3):***415
nonpetroleum hydrocarbons ***5(2):***7
NTA, *see* nitrilotriacetate
nutrient(s) ***5(1):***283; ***5(3):***201, 265, 505
nutrient addition ***5(3):***383, 499

O

O_2, *see* oxygen
o-chloroaniline, *see* chloroaniline
octane enhancer ***5(3):***25
oil ***5(6):***51
oil spill(s), *see* crude oil *and* fuel oil
oil well blowout, *see* crude oil
oily sludge ***5(3):***215
on-line monitoring ***5(3):***517
on-site disposal ***5(6):***259
optimization ***5(2):***275
ORC, *see* oxygen-release compound
organic acid(s) ***5(1):***343
organic compound(s) ***5(6):***15
organic matter ***5(3):***463; ***5(8):***179
organomercurial ***5(3):***433
organophosphate(s) ***5(6):***63
OXC, *see* oxidation capacity
oxidation ***5(1):***59; ***5(3):***209, 463
oxidation capacity (OXC) ***5(1):***189
oxidation/reduction potential ***5(1):***325, 343
oxidative coupling ***5(5):***189
oxygen (O_2) ***5(1):***147; ***5(3):***13; ***5(5):***31; ***5(7):***223
oxygen-release compound (ORC) ***5(2):***15; ***5(3):***13, 271, 277, 283, 289, 295, 301, 319, 427; ***5(6):***169; ***5(7):***223
oxygen transfer ***5(3):***103
oxygenate(s) ***5(3):***7; ***5(5):***129
ozone ***5(5):***25, 203

P

P. chrysosporium, see Phanerochaete chrysosporium
P. fluorescens, see Pseudomonas fluorescens
P. ostreatus, see Pleurotus ostreatus

packed-bed reactor ***5(5):***161
PAH(s), *see* polycyclic aromatic hydrocarbon(s)
PAH-ketones ***5(8):***135
PAH-quinones ***5(8):***135
partitioning ***5(8):***271
passive treatment ***5(6):***217
Pb, *see* lead
PCB(s), *see* polychlorinated biphenyl(s)
PCE, *see* perchloroethene and tetrachloroethene
PCP, *see* pentachlorophenol
PCR, *see* polymerase chain reaction
peat ***5(2):***141
pentachlorophenol (PCP) ***5(5):***97; ***5(7):***75, 83, 89, 229
pentane ***5(3):***31
perchlorate ***5(4):***147; ***5(6):***239
perchloroethene (PCE) (*see* tetrachloroethene)
permeability ***5(2):***249; ***5(6):***175
permeable treatment walls ***5(2):***237
persulfate oxidation ***5(8):***241
pesticide ***5(5):***203; ***5(6):***151
petroleum *see* crude oil *and* fuel oil
petroleum hydrocarbon(s) ***5(1):***47, 141, 153, 247; ***5(3):***59, 65, 89, 131, 137, 143, 169, 195, 209, 265, 271, 319, 325, 331, 345, 375, 389, 395, 415, 427, 439, 445, 487, 499, 505, 517, 523, 541, 545; ***5(5):***31, 143, 155, 57, 75, 81, 177; ***5(6):***27; ***5(8):***63
PFBR, *see* pressurized fluidized bed reactor
pH ***5(2):***191
Phanerochaete chrysosporium ***5(5):***103; ***5(7):***173; ***5(8):***93
phenanthrene ***5(3):***389; ***5(8):***191, 247
phenols ***5(1):***225, 277; ***5(8):***13
phospholipid(s) ***5(1):***171
phospholipid fatty acid (PLFA) ***5(3):***345, 545
photolysis ***5(7):***161
phreatophyte ***5(6):***21
phthalates ***5(6):***187
physico-chemical processes ***5(7):***191
phytoremediation ***5(6):***9, 21, 27, 33, 45, 51, 95, 101, 115, 121, 127, 133, 139, 151, 239; ***5(7):***33, 191
phytotoxicity ***5(5):***13; ***5(6):***51; ***5(7):***39
plant uptake ***5(6):***127
plants ***5(6):***145
Pleurotus ***5(7):***69
Pleurotus ostreatus ***5(8):***93, 99
PLFA, *see* phospholipid fatty acid
plug-flow reactor ***5(8):***1
plumes ***5(1):***97
polychlorinated biphenyl(s) [PCB(s)] ***5(6):***9; ***5(7):***149, 155, 161, 167, 173, 179, 185, 241
polycyclic aromatic hydrocarbon(s) [PAH(s)] ***5(1):***265, 289; ***5(3):***337; ***5(5):***1, 13, 25, 37, 69, 87, 97; ***5(6):*** 9, 39, 45, 51, 69, 75; ***5(8):***1, 7, 13, 19, 25, 31, 37, 43, 51, 57, 63, 69, 81, 87, 93, 99, 105, 111, 117, 123, 129, 135, 141, 147, 155, 161, 167, 173, 179, 185, 191, 197, 203, 217, 223, 229, 235, 241, 247, 253, 265, 271, 283, 289, 295, 301, 309, 315
polylactate esters ***5(2):***287
polymerase chain reaction (PCR) ***5(1):***1; ***5(3):***357, 363
poplar tree ***5(6):***133
population dynamics ***5(4):***1; ***5(6):***277
potassium permanganate ***5(2):***199
precipitation ***5(4):***121
prediction ***5(7):***185; ***5(8):***253
preoxidation ***5(6):***157
pressurized fluidized bed reactor (PFBR) ***5(2):***217
pristane ***5(3):***389
productivity ***5(8):***295
propane ***5(2):***1, 95
Pseudomonas ***5(1):***171; ***5(3):***535
Pseudomonas aeruginosa ***5(3):***31
Pseudomonas fluorescens ***5(3):***433
public-private partnership ***5(6):***295
pulp mill wastewater ***5(6):***157
pulse venting ***5(3):***183
pumps ***5(3):***143
pure oxygen injection ***5(3):***65
push-pull test ***5(3):***83, 109, 123
pyrene ***5(8):***75

Q

quinones ***5(8):***75

R

Ra, *see* radium
radiation ***5(5):***63
radioactive isotopes ***5(4):***103
radio-respirometry ***5(3):***545
radiowave ***5(3):***487
radium (Ra) ***5(4):***103
radius of influence ***5(3):***103
Raoult's law ***5(3):***19
rates, *see* kinetics
RBCA, *see* Risk-Based Corrective Action
RDX ***5(7):***1, 15, 21, 33, 45, 57, 217
reactive barrier, *see* reactive wall
reactive wall ***5(2):***205; ***5(3):***541; ***5(4):***19
recirculation ***5(6):***251
redox, *see* oxidation/reduction potential
reduction ***5(4):***79, 121, 135
reductive dechlorination ***5(1):***1, 53, 83, 89, 253; ***5(2):***35, 47, 55, 67, 73, 121, 129, 147, 165, 171,181, 185, 225; ***5(6):***101; ***5(7):***259
reductive transformation ***5(7):***95
reeds ***5(6):***109
regulatory acceptance ***5(1):***355
regulatory strategy ***5(1):***295
remote monitoring ***5(6):***259
renewable energy ***5(5):***149
resistive heating ***5(3):***403
resource damage ***5(6):***289
respiration ***5(1):***153; ***5(3):***109, 183
respiration test ***5(3):***109
respirometry ***5(3):***115; ***5(5):***183
reverse osmosis ***5(4):***97
rhizodegradation ***5(6):***239
rhizosphere ***5(6):***27, 33
risk assessment ***5(1):***359, ***5(6):***231, 283
Risk-Based Corrective Action (RBCA) ***5(1):***219, 349
risk-based design ***5(1):***349
risks ***5(8):***57
RT3D ***5(2):***165
Ruelene™ ***5(6):***151

S

salt tolerance ***5(6):***21
sand/sawdust ***5(2):***237
sanitary landfill ***5(6):***181
sediment(s) ***5(6):***39, 187, ***5(7):***95, 167; ***5(8):*** 31, 223, 271, 283, 289, 309
selenium ***5(4):***141
semiarid ***5(1):***89
semifield scale ***5(3):***529
semiquinone ***5(8):***75
septic tank ***5(4):***85
SEQUENCE, *see* visualization tool(s)
sequencing batch reactor, *see* batch reactor
sequestration ***5(8):***283, 289
sewage sludge ***5(6):***15
SF6, *see* sulfurhexafluoride
signature lipid biomarkers ***5(4):***1
site management ***5(6):***283
sludge ***5(5):***37
sludge stabilization ***5(6):***277
slurry ***5(7):***33, 179; ***5(8):***1
slurry reactors ***5(6):***75
SOBs, *see* sulfur-oxidizing bacteria
soil heating ***5(3):***487
soil moisture ***5(3):***201
soil productivity ***5(5):***189
soil resistance ***5(6):***175
soil shearing ***5(2):***249
soil vapor extraction (SVE) ***5(3):***19
soil washing ***5(5):***19; ***5(7):***101, 161
solid waste ***5(6):***181
solid-phase microextraction (SPME) ***5(3):***71
solid-state bioreactor ***5(5):***155
solvents ***5(1):***89
sparging, *see* air sparging
species richness ***5(6):***33
SPME, *see* solid-phase microextraction
16SrDNA ***5(5):***143
stability ***5(4):***115
stable isotope(s) ***5(1):***207; ***5(5):***183
starvation ***5(4):***13
starved cells ***5(2):***211
statistical study ***5(1):***337
styrene ***5(5):***161
sulfate ***5(4):***127
sulfate reduction ***5(2):***165; ***5(4):***19
sulfate-reducing bacteria ***5(6):***193, 205, 211, 225; ***5(7):***235
sulfide ***5(8):***309
sulfurhexafluoride ***5(3):***123
sulfur/limestone ***5(4):***85
sulfur-oxidizing bacteria (SOBs) ***5(4):***127
sulphate, *see* sulfate

sulphide, *see* sulfide
surface hydrophobicity ***5(6):***163
surfactants ***5(5):***167, 177; ***5(7):***155, 161; ***5(8):***81
sustainable remediation ***5(6):***283
SVE, *see* soil vapor extraction

T

T. versicolor, *see Trametes versicolor*
tannery wastewater ***5(6):***157
TBA, *see tert*-butyl alcohol
TCA, *see* trichloroethane
1,1,1-TCA, *see* 1,1,1-trichloroethane
TCE, *see* trichloroethene and trichloroethylene
temperature, *see* cold temperature
tert-butanol ***5(3):***25
tert-butyl alcohol (TBA) ***5(1):***103
tetrachloroethene (PCE) ***5(1):***7, 29, 253; ***5(2):***21, 27, 47, 55, 121, 129, 141, 181, 205, 225
Th, *see* thorium
thermodynamic ***5(1):***201
thermodynamic model ***5(6):***163
thermophilic ***5(6):***277
Thiessen Polygon Method ***5(1):***141
thorium (Th) ***5(4):***103
tides ***5(1):***147
TNB, *see* trinitrobenzene
TNT, *see* 2,4,6-trinitrotoluene
toluene ***5(1):***189, 207; ***5(3):***71, 469; ***5(5):***117, 123
toluene ortho-monooxygenase ***5(2):***293
total petroleum hydrocarbons (TPH, TPHC) ***5(1):***153; ***5(3):***277; ***5(5):*** 7, 13, 37, 43, 87, 167; ***5(6):***51, 63; ***5(8):***309
town gas ***5(8):***7, 123
toxaphene ***5(6):***89; ***5(7):***125
toxicity ***5(3):***351, 369, 389, 395; ***5(4):***7, ***5(5):***37, ***5(7):***229, ***5(8):***185, 217
TPH, *see* total petroleum hydrocarbons
TPHC, *see* total petroleum hydrocarbons
Trametes versicolor ***5(8):***93
transformer oil ***5(7):***149
translocation ***5(6):***127
transport ***5(3):***433
transport model ***5(2):***27
trichloroethane (TCA) ***5(1):***259
1,1,1-trichloroethane (1,1,1-TCA) ***5(1):***71
trichloroethene and trichloroethylene (TCE) ***5(1):***1, 41, 47, 65, 77, 241; ***5(2):*** 89, 95, 101, 107, 121, 141, 147, 157, 181, 191, 199, 217, 225, 263, 275, 293, ***5(5):***161; ***5(6):***101, 127
trinitrobenzene (TNB) ***5(7):***33
2,4,6-trinitrotoluene (TNT) ***5(6):***9; ***5(7):*** 27, 33, 39, 45, 57, 63, 197, 203, 211
two-phase extraction ***5(3):***155

U

uranium ***5(4):***103, 115; ***5(7):***235

V

vacuum ***5(3):***143
vacuum-enhanced ***5(3):***163
vacuum-enhanced free-product recovery ***5(3):***149
vadose zone ***5(1):***89, 213, 219, 225, 231; ***5(7):***241
vapor phase ***5(5):***123, 135
vapor-phase bioreactor ***5(5):***129, 135
vapor-phase transport ***5(1):***219
vegetable oil ***5(4):***47, 53
vinyl chloride ***5(1):***77; ***5(2):***15, 81
viral sorption ***5(6):***163
visualization tool(s) ***5(1):***325
vitamin B_{12} ***5(2):***205
VOC(s), *see* volatile organic compound(s)
volatile organic compound(s) [VOC(s)] ***5(5):***135
volatilization ***5(4):***141
volatilization-to-biodegradation ratio ***5(3):***195

W

waste gas ***5(5):***161
wastewater ***5(5):***203
weathered crude oil ***5(8):***51
wetland ***5(3):***325; ***5(6):***57
white-rot fungi ***5(6):***271; ***5(7):***33, 69; ***5(8):***69, 81, 87, 93, 99, 185
wood preservative(s) ***5(5):***97; ***5(7):***89; ***5(8):***167

Z

zero-valent iron ***5(2):***205
zinc (Zn) ***5(4):***25; ***5(6):***217
Zn, *see* zinc